Fleece & Fibre

Textile Producers of Vancouver Island and the Gulf Islands

Fleece & Fibre

Francine McCabe

Copyright © 2023 Francine R. McCabe

All rights reserved. No part of this publication may be reproduced, stored in a retrieval system, or transmitted in any form or by any means—electronic, mechanical, audio recording, or otherwise—without the written permission of the publisher or a licence from Access Copyright, Toronto, Canada.

Heritage House Publishing Company Ltd.
heritagehouse.ca

Cataloguing information available from Library and Archives Canada
978-1-77203-453-0 (paperback)
978-1-77203-454-7 (e-book)

Edited by Marial Shea
Proofread by Renée Layberry
Cover and interior design by Setareh Ashrafologhalai
Cover photographs, by Francine R. McCabe: An Angora kid under its mother at Up A Creek Farm (*front*); Woven Hemp at Heritage Fibre Mill (*back*). All interior images are by the author unless otherwise indicated.

Image credits:
p. i, Indigo plants growing in the garden at Hinterland Yarns; *p. ii–iii*, Garments; *p. iv–v*, Antique wool shearing and processing tools at Checker Grass Farms; *p. viii–ix*, The lead border collie at Guthrie Farms bringing in the flock of Shetland sheep; *p. xiii*, Cross-bred lamb at Genesta Farm; *p. xiv*, A llama at Millstream Miniature Llamas; *p. 19*, Shorn fibre at Genesta Farm; *p. 159*, Flax seed heads harvested from the one-acre planting in Saanichton; *p. 206-7*, Babydoll Southdowns at Furrycreek Farm; and *p. 208*, A llama at Millstream Miniature Llamas.

The interior of this book was produced on FSC®-certified, acid-free paper, processed chlorine free, and printed with vegetable-based inks.

Heritage House gratefully acknowledges that the land on which we live and work is within the traditional territories of the Lekwungen (Esquimalt and Songhees), Malahat, Pacheedaht, Scia'new, T'Sou-ke, and W̱SÁNEĆ (Pauquachin, Tsartlip, Tsawout, Tseycum) Peoples.

We acknowledge the financial support of the Government of Canada through the Canada Book Fund (CBF) and the Canada Council for the Arts, and the Province of British Columbia through the British Columbia Arts Council and the Book Publishing Tax Credit.

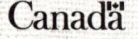 

27 26 25 24 23   1 2 3 4 5

Printed in China

For the farmers:
Your commitment to the land,
your animals, and
your crops is invaluable.

Contents

Introduction:
In Search of a Fibreshed   1

PART I
Animal-Based Fibres   19

1 | Sheep   23

2 | Alpaca   121

3 | Llama   133

4 | Angora Rabbit   143

5 | Goat   147

PART II
Cellulose-Based Fibres 159

6 Flax 167

7 Hemp 177

8 Invasive Species 183

Conclusion 191

Resources 197

References 201

Acknowledgements 203

"To love a place is not enough. We must find ways to heal it."

ROBIN WALL KIMMERER

Introduction: In Search of a Fibreshed

◂ Clean Shetland fibre, ready to card or spin as is from Guthrie Farm.

THE WOOL BLANKET draped over the couch, the cashmere sweater warming your shoulders, the cotton bedding you'll climb into tonight—they all started from the ground somewhere. Do you ever wonder where the fibre grew and how it was processed to become the useful item it is today? *Fleece & Fibre: Textile Producers of Vancouver Island and the Gulf Islands* was born out of my curiosity to find the islands' fibre. I wanted to know what we had here, who was growing it, how, and why. And as a spinner and weaver, I wanted to use only local fibre in my weavings. So I set out on a search for island fibre and fell in love with our flourishing fibre economy.

My journey began in early fall, on one of those days on the island when the fog is so thick the land disappears under a smothering cloud. My family and I were out on a weekend adventure, driving up the Old Island Highway to explore a new area. Once we were out of the fog, I counted several farms with sheep, goats, and acres being readied for the coming season. It seemed obvious to me that the island would be producing its own fibre. After all, I had been to fibre festivals before Covid and saw lots of fibre there. But I knew by the number of farms I could see on the drive that there had to be more. For a while, I'd been thinking about making something with only island-grown fibre. Now, as I was dreaming up a new weaving, the beginning of my fibre search was sparked.

Later that week, I visited a few of my favourite yarn shops and discovered that they had little to no local fibre. When I asked the

▲ Hinterland Watershed yarn.

▶ Mixed flock of sheep waiting to be sheared at Parry Bay Farm.

shop staff why, they all told me the same thing. By the time the farmer has the fibre processed and packaged into yarn, the cost is already high. Once the shop adds enough markup to make a profit, the price is beyond what most consumers will pay.

This left me wondering: Why was local fibre so costly to produce?

I discovered that Vancouver Island once had several small family-run fibre processing mills, but I couldn't track down any existing mills. Today, farmers looking to process their fibre have no other option than to ship their fibre off the island, sometimes even across the country, for affordable processing. With the huge variety of fleece and fibre on the island, both animal- and plant-based, I was shocked to find no operating mills.

The problem was certainly not a lack of fibre. The more I searched for local fibre, the more amazed I was by the variety of breed-specific producers. I began by compiling a list of farms from organizations like the BC Sheep Federation, Wool.ca, and the Vancouver Island Llama and Alpaca Club. Each resource led to another, and soon I had a long list of farms that were breeding fibre animals.

Yet with all this fibre being produced, I still didn't know where it was all going or how I could get my hands on it. And I *really* wanted to get my hands on it! So, I did the only other thing I could think of: I started knocking on doors and contacting people by phone or email.

Early on, I came across a compelling concept that would inspire, inform, and ground my search for a local fibre economy.

Discovering the Fibreshed Concept

The fibreshed concept was pioneered by Rebecca Burgess, M.Ed., executive director of the California-based non-profit organization Fibershed. (You'll notice two different spellings: "fibre" in Canada and "fiber" in the US.) She defines "fibreshed" as "a vision that

▶ A pile of wool that was advertised as free. The farmer was going to bury the rest that wasn't taken.

enhances social, economic, and political opportunities for communities to define and create their fiber and dye systems and redesign the global textile process. It is place-based textile sovereignty, which aims to include rather than exclude all the people, plants, animals, and cultural practices that compose and define a specific geography." (Learn more from Burgess's book, *Fibershed: Growing a Movement of Farmers, Fashion Activists, and Makers for a New Textile Economy*.)

Burgess's concept made so much sense to me. I realized I was far from the first person to be asking these questions about local fibre and the growth of a regional textile economy. Her concept has quickly spread, and networks have begun popping up all over the world. Rich in fibre resources, Vancouver Island and the surrounding Gulf Islands incorporated in 2018 into the fibreshed movement.

As I started to connect with my local fibreshed, I learned about its efforts to educate the public and grow the connection between farmers and fibre consumers. They held events like Getting Value or Getting Fleeced, with a panel of local experts in agriculture to educate farmers on maximizing the value of their fleece. They have an active producer page and promote their producers through a regular newsletter and social media.

Gaps in the System

I spent over a year sourcing island fibre and visiting farmers and their fibre animals. Along the way, I kept hearing similar stories of processing woes and the growing cost of feeding and caring for animals. This made me wonder: Why aren't we, as a province, investing more in fibre infrastructure? Why are there multiple agricultural grants for farmers looking to produce food, but so little financial help for starting fibre-related farms? We need sustainable start-up options for fibre mills. Closing this gap is one of the goals of the fibreshed movement. Farmers need an easy way to turn their

fibre over so they can justify and afford the time spent on the product, and the makers that are looking for local fibre need to know what is locally available and how they can buy it. All the parts are here; they just need connection.

Farmers themselves, however, are working hard to close gaps. Each farmer in this book made it clear that using all the materials from their farm in a renewable way is important to them and their farm practices. I saw fields of flax used as a dual-purpose cover crop. I met farmers with fibre mill plans in the works. One farmer told me about a new wool pelletizer she just purchased. Others are using their waste wool to build road stability.

◀ Freshly shorn fibre at Parry Bay Farm.

▲ Babydoll Southdown ewe grazing at Yellow Point Farms' seasonal petting farm.

What these farmers are doing to care for their animals, crops, and the land is invaluable and deserves to be shown off. And the products they are bringing to our local market are worth supporting and nurturing. Everyone I met along the way has been so enthusiastic about fibre and the possibilities. They give me hope.

Why Our Material Choices Matter

The way we shop and care for our clothing can support a local fibre system and contribute to the regeneration of our soil. Built on greed and profit, our current textile and garment industries are outdated. We need a new system that considers the wellness of animals, people, and the environment at every level of production. Never before has clothing been so accessible and affordable. High-end knockoffs and ready-to-wear, up-to-the-minute fashion trends are flowing out so fast that consumers can't keep up. How did we get here?

Twentieth-century globalization led to a drastic change in the garment industry through the gradual introduction of mass production and fast fashion. Production was moved overseas, where cheap labour markets were exploited. When online shopping exploded in the 2000s, fashion trends accelerated, and clothing was pumped out faster and in larger numbers than ever before.

But this convenience came with a price. According to the Government of Canada, "textiles account for approximately 6% of plastic use and waste in Canada for products such as clothing, carpets, footwear, fabrics and upholstery." Thanks to the garment industry's shedding of microfibres, it has become one of the largest polluters of our world's water supply at every step of the way, from production to laundering and disposal.

According to the United Nations Environmental Programme (UNEP) and the Ellen MacArthur Foundation, "Every year a half a million tons of plastic microfibers are dumped into the ocean, the equivalent of 50 billion plastic bottles." These plastics are so small it is impossible to extract them, which leads to devasting effects on animal and human health.

Microplastics are only one of the major environmental problems created by fast fashion. The production and transportation of our textiles is contributing 10 percent of annual global carbon emissions. As well, our overflowing landfills are increasing by the second and the pesticide-heavy crops are adding to the pollution of our land and water, along with the numerous labour violations associated with the global industry.

To change this system, we must begin with ourselves. We cannot avoid the amount of interaction we have with textiles; the world would be hard and cold without them and we would lose an important means of self-expression. But we can make choices that will support change.

We no longer know what is in our textiles. We need to demand the same transparency from clothing brands as we have with our food. Even products labelled as being 100 percent natural fibres are often finished with some sort of chemical treatment, leaving them too toxic to return to the soil. These are all transparency issues that need to be addressed by brands. Good questions to ask include: Where was this product made and by whom? Is the brand that produced the product using any transparency in its production? Is it a product that is made to last?

We can also work toward change by becoming mindful of our shopping habits and what compels us to purchase. I'm not saying don't buy something new, but do consider a few questions first. Do you need another shirt when you have one at home that can be mended? How does this new garment fit into your existing wardrobe? Will you get long-term use out of it? Can you define your sense of value? Does it come from the latest trends, or are you someone who appreciates the story behind a garment?

The clothing and textiles we choose speak to our values. We can choose to make, mend, re-wear, re-purpose, tailor, thrift, and trade all while still being unique and fashionable. We might think that people who thrift, mend, re-wear, and repair clothing cannot afford new clothing, but this is the kind of cultural conditioning fast fashion thrives on. Wearing clothing for years and repairing and caring for it is something to be proud of. We can make a statement about our principles and help our planet all while staying stylish. And as a bonus, garments that are mended are loved and made unique in the process.

Not only do we have more fashion options than ever before, we also have the information to educate ourselves on the global textile industry and its devastating environmental, social, and human health costs. We can go back and reset our wardrobes, letting them be a tactile rebellion against fast fashion. Let's slow down and bring our textiles home.

I encourage fellow fibre artists to source their fibre locally. Even the fibre that isn't good for making next-to-skin textiles can make amazing wall art, rugs, fibre installations, dryer balls, dish scrubbies, and other durable textile products we use daily around our homes.

Vancouver Island and the Gulf Islands have many brands creating from the depths of our land and consciously working to better the local textile options. We are lucky to be in a region where we are rich not only in fibre, but in people who care about taking the steps we need to change how we produce and consume textiles and fashion. We have the materials, the makers, and the market to build a unique island textile industry.

A Fibreshed in Search of Infrastructure

Vancouver Island has nearly all the moving parts needed to call ourselves a fully functioning fibreshed. We grow a large variety of animal and plant-based fibre material and we have a huge network of people interested in processing those raw materials into finished products. And we have the makers and small businesses looking to use local materials in the production of their products. But our

▶ Raw Shetland fleece from Root Spell Shetlands.

infrastructure for processing those fibres at a commercial level is lacking.

Residents of Vancouver Island and the Gulf Islands pride themselves on the wide selection of local, organic, and sustainably grown food. As a region, we have put a lot of effort and time into ensuring we have access to high-quality foods because we know how important it is to put healthy products into our bodies that also help our local economy thrive. But it seems the production of our plant and animal-based fibres and dyes for making our clothing and textiles has been an afterthought.

Fortunately, we are beginning to realize that our textiles are just as impactful to our region as the food we eat. We engage with clothing and textiles just as often as food, if not more. We sleep in them, we wear them all day, we use them to furnish our homes; they are everywhere.

Fibre and food start the same way—from the ground up. Fibre and clothing are both agricultural products that start with our dependence on the land and farmers. Just because we are not consuming our clothing doesn't mean we aren't absorbing them. And our land, in turn, must digest these unnatural fibres and chemicals once we throw them out.

A 2018 survey conducted by Dalhousie University asked 1,046 people to fill out an online questionnaire regarding genetically modified ingredients in their food. Close to 90 percent of people surveyed said there should be mandatory labelling of GMOs. So why are we willing to put clothing on the largest, most absorbent organ of our bodies without even questioning why it has no ingredient label? The multitude of chemicals and synthetic fibres used to process our clothing is a large contributor to greenhouse gases, not to mention the undocumented health issues it causes.

Genetic engineering of our textiles has been the high-tech solution offered to address the major issues we have created with our outdated, fabricated, and toxic production. The introduction of new technologies and biosynthetic fabrics, like spider-silk, are being pushed at us as if they are the "natural" solution to the massive amount of waste and destruction our current textile economy has created. But when broken down, these biosynthetic products depend on critical products like sugar, whose production has also been linked to deforestation and poor labour conditions. These unsustainable practices are diverting support away from the growth of viable regional fibre economies. We are so used to overconsumption that most of us do not even question these technologies.

If you went to your closet, could you find a piece of clothing that is entirely organic and would fully break down in a compost pile and benefit our soil? Do any of the clothes even have a label that lists all the materials and chemicals used to produce them? How about the amount of water they took to produce? My wardrobe certainly doesn't have such a garment.

I heard someone say, "I once worked in a textile factory and I wouldn't want my children working in one of those." But the mills and infrastructure of today are not the factories of the past. Mills operating in other provinces, like Custom Woolen Mills in Alberta and Longway Homestead in Manitoba, are family-run operations that take raw fibre and process it into roving, batts, and yarns, all without chemicals and in a way that contributes to their economy.

We are lucky enough to live in a region that has a multitude of fibre sources as well as people looking to turn that fibre into useable textiles. However, our fibre is sitting in warehouses, sometimes for years, before being processed at the busy mills. We don't have the local infrastructure we need to process those fibres at a commercial level. Farmers shared with me that the cost of shipping their fibre to the mills is expensive. With no promise on return time or a definite buyer of the finished product, they often find it easier to sell the raw fleece, give it away, compost, or dispose of it. Some farmers were lucky and knew people travelling to and from cities with mills and could send their fibre along with them, saving them the cost of shipping. But even then, they would still have to market the product somehow, and that is a time-consuming challenge for already busy farmers.

Here on the islands, one of the main ways farmers have connected with makers is through annual fibre festivals and fairs. When Covid hit, all fleece and fibre fairs were cancelled for a time and the connection between makers and the raw product got a lot more distant. I have noticed that farmers, designers, makers, and fibre enthusiasts alike are now trying to create more connections through Facebook groups and online communities, perhaps in response to the pandemic.

The potential for a textile industry is here. We have the product. Now we need the infrastructure. We could create more jobs as well as open up economically viable options for farmers looking to process their fibres. And we are also seeing a larger, cross-Canada fibreshed movement growing more connections and resources toward a national infrastructure (see canadianfibreshed.org).

We can bring our fibres home, back to the soil, not just to benefit those able to afford handmade clothing, but to build a Canadian textile economy. This would create jobs and decrease production costs with demand and supply. It would open more opportunities for producers, designers, and makers to start their own locally sourced businesses. Consumers would have more transparent options for their clothing and textiles. And we could contribute to the regeneration of our soil.

The possibilities are infinite, and I could go on and on. Before you buy your next garment, I urge you to educate yourself, think deeply about the impacts of fast fashion, and turn instead to some of the wonderful fibre producers in the following pages.

How to Use this Book

I have organized this book by fibre types. For example, if you are looking for sheep wool, you can flip to that section and find

▲ The fences at Reynolds Family Farm are covered with cashmere fibre from the goats rubbing their bodies along the fence line.

▲ Salish Woolly Dog. Ian McTaggart Cowan Collection. UNIVERSITY OF VICTORIA SPECIAL COLLECTIONS / W̱SÁNEĆ LEADERSHIP COUNCIL

information about the breed types available locally, and who is selling the raw fleece, processed fibre, and finished products. Some fibre farmers are cross-breeding to get the best quality and quantity of fleece and protein from their flock; they are listed alphabetically, by breed.

I have also indicated what each type of fibre is good for from a maker's point of view. For example, some breeds are best for making garments while others are more suited for making durable textiles. Some fibres can be spun fine, while others will felt.

I contacted over one hundred farms, and would have loved to visit every single one, but it wasn't doable. This is because some farmers were no longer farming or doing anything with fibre, or I wasn't able to connect with them, while others were too busy or didn't want to be included, and a few practised strict biosecurity, not allowing visitors. Some farmers I interviewed by phone. Given this difference in access, some farms have a larger write-up than others.

I hope this book will give you insight into our regional fibre economy. My goal is to help bridge the gap between the farmers growing the product, the makers who produce the fibre into wearable, usable items, and consumers looking to support the local textile industry.

Direct contact information is not provided for privacy purposes. Go online to your favourite search engine and find websites and Facebook or Instagram pages listing a farm or organization's most recent products and contact information.

PART I
Animal-Based Fibres

ANIMAL-BASED FIBRES include any fleece or fibre harvested from an animal for human use. The most common domesticated fibre animals on Vancouver Island and the Gulf Islands are sheep, alpacas, llamas, goats, and rabbits.

Sheep (*Ovis aries*) are the predominant protein fibre in our region. Humans have benefitted from sheep fibre for millennia. But before sheep wool became the popular fibre here, Coast Salish Peoples used other fibres.

The Coast Salish Peoples were well-established wool workers well before European settlers made their way to the West. They kept woolly dogs (now extinct), with pointed ears, curled tails, and long, thick coats. Dr. Elaine Humphrey and her team at the University of Victoria's Advanced Microscopy Facility have studied over fifty-one blankets from museums and private collections confirming the use of dog fibre in Coast Salish textiles.

Woolly dogs lived on what are now known as the Gulf Islands and other small surrounding islands, so they could not cross-breed with other domesticated dogs. During the early spring months, the dogs' fleece would be at its thickest. Coast Salish women would wash the dogs and hand shear them with sharp stones or shell knives. These dogs grew enough fibre to be sheared or trimmed up to three times a year. Terrence Loychuk, a private researcher and collector working with Dr. Humphrey, speculates that the fibre may have simply shed from the dogs, similar to the coat of a Samoyed, rather than needing to be cut off.

However, Coast Salish people did not use the dog wool alone. They combined it with mountain goat wool to create a stronger, longer-lasting fleece for spinning and making into blankets, rugs, shelters, garments, and other useful textiles. Mountain goats are not native to Vancouver Island, so people would trade with neighbours from the mainland. Minnie Peters, a Sto:lo weaver, explains in a YouTube video that "the warriors used to go up the mountain to hunt, and they used to see the white material on the bushes, and they gathered that and brought it down to their wives." She goes on to explain that the fibre wasn't good for spinning on its own. So the mainland Coast Salish people would then trade with the Island Coast Salish for the dog fleece. On occasion, they would hunt the mountain goats but the meat was not a desired protein and the hunt was often too dangerous as the goats would retreat to higher and higher ground.

Dog and mountain goat wool provided the woolworkers with the base fibre, but they often used plants like stinging nettle, Indian hemp, milkweed, cottonwood, fireweed fluff, cedar twine, and feathers from waterfowl to bulk up their wool. Dr. Humphrey explains that "the plant fibre found under the microscope often helps [the research team] determine the other fibres."

By the late nineteenth century, the increasing presence of European settlers led to the decimation of the woolly dog and the decreased need for mountain goat fibre as sheep were introduced.

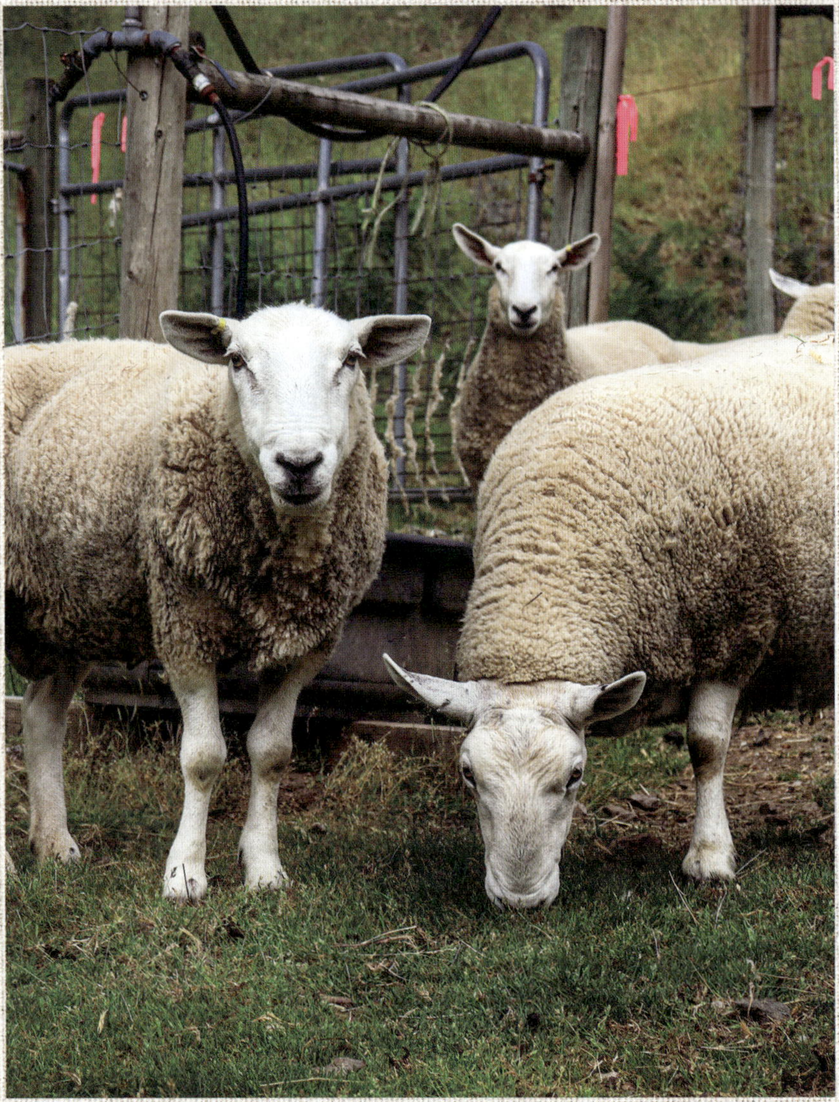

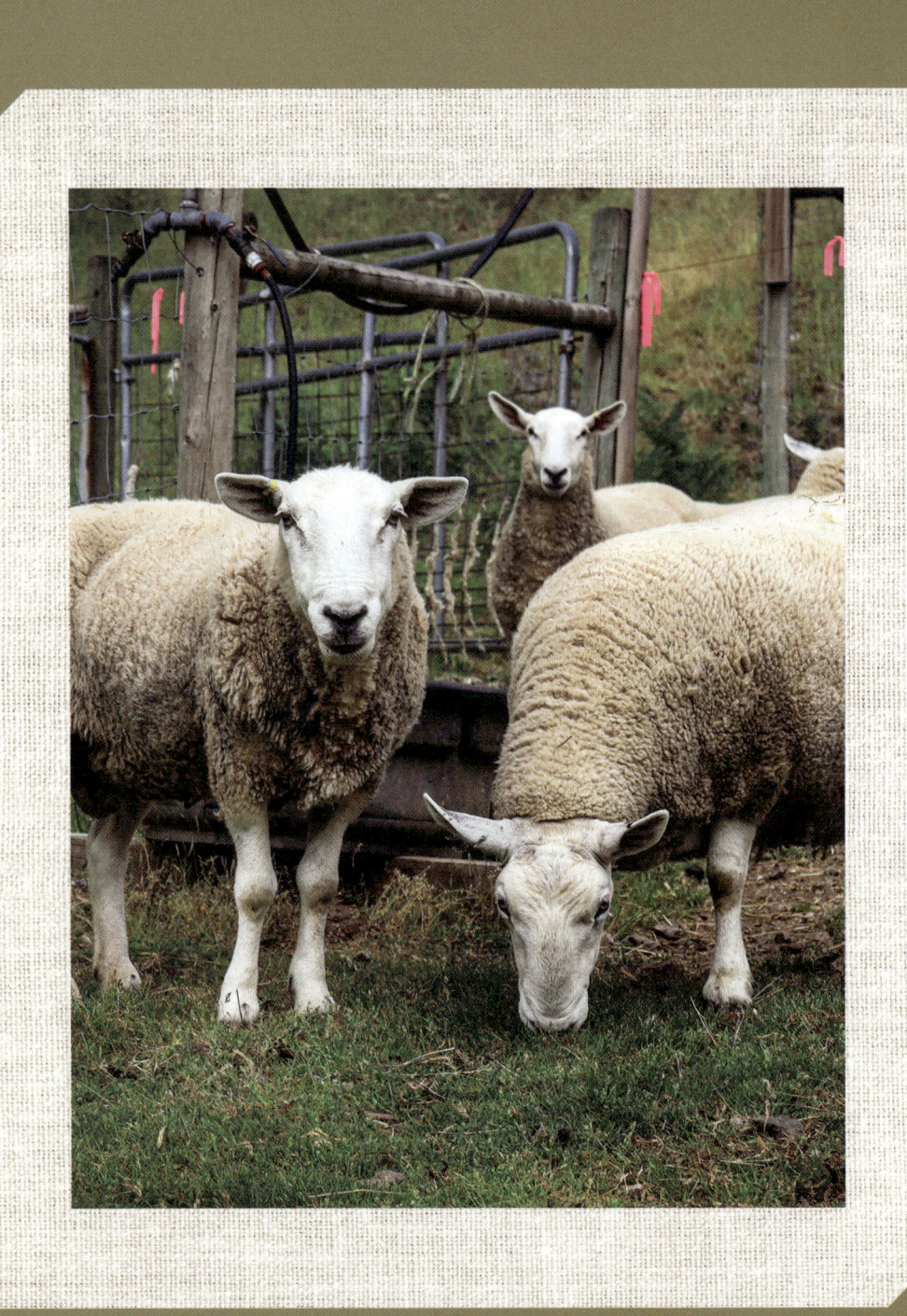

1 Sheep

◀ North County Cheviots at Conheath Farm.

The Wonders of Wool

Wool has many unique properties, several of which make it one of the most comfortable and versatile fibres to wear. Because wool can absorb up to 30 percent of its weight in moisture from the air around it and the wearer's perspiration, it is ideal for our damp winter climate. As well, it is a temperature regulator, not an insulator; this means that when water enters the fibre, it creates a temporary chemical reaction that creates energy, allowing the material to stay warm when wet and keep us cool in hot, dry weather.

Though some people claim to be unable to wear wool, the fibre is actually non-allergenic. Those who react to it may be more sensitive to its scaly surface. They could also be reacting to a chemical treatment that has been applied to the wool, such as mothproofing, a common treatment among many that are harmful to our skin and environment.

Wool has many safety features that make it practical. At a microscopic level, it is covered in a layer of scales that help it resist dirt and dust. The fibre is also considered fire resistant; though it will burn, it takes a while to get going, and it will self-extinguish when the flame is removed.

Finally, another major benefit of wool is that, if processed without chemicals, it can biodegrade and add nutrients back to the soil.

A Brief History of Wool in Our Fibreshed

By 1850, sheep were the predominant fleece-bearing animal on Vancouver Island. With the arrival of European settlers, Coast Salish Peoples were introduced to knitting, a post-contact acculturative that would have come naturally to the already-skilled fibre workers.

One product most people are familiar with is the Cowichan sweater. What could be more iconic to our island than this durable, hard-wearing sweater hand-knit from bulky-weight yarn—a garment that starts on this land, is made here by the people of this land, and tells the stories of this land?

For a time, Cowichan sweaters were worn only by Indigenous Peoples. But by the 1920s, the sweater became popular, and Coast Salish knitters felt the pressure to keep up with demand. In the 1960s, one knitter, Sarah Modeste, and her husband decided to invest in a commercial carder that fast-tracked the wool processing and made roving for the Cowichan spinners in half the time. Although the carder helped to speed up production, overseas interest in the product was overrunning the market, and soon knock-offs were being produced.

However, the traditional way of making a Cowichan sweater cannot be replicated. Each sweater is unique and knit in a way that leaves it seamless besides the shoulders. The designs knit into the sweaters are individual to each family of knitters and passed down through the generations.

In 2011, the Coast Salish knitters and the Cowichan sweater were recognized by the Canadian government as historically significant and designated as a National Historical Event by the Historic Sites and Monuments Board of Canada. To protect the authenticity of their textiles, the Cowichan knitters registered "Cowichan," "Genuine Cowichan," and "Genuine Cowichan Approved" as trademarks for garments made "with the traditional tribal methods by members of the Coast Salish Nation using raw, unprocessed, undyed, hand-spun wool, also made and prepared in accordance with traditional tribal methods."

Today, Cowichan knitters are still designing and making Cowichan sweaters, ponchos, vests, mitts, head wear, footwear, scarves, and accessories. Registered knitters are listed on the Cowichan Tribes website with other entrepreneurs in Arts and Crafts.

At one time, the Cowichan knitters and spinners were the largest purchasers of island wool. Many farms shared with me that the same buyers came back year after year until they didn't. The farmers, when asked, weren't sure why the Cowichan knitters stopped buying from them.

One article in *YAM* magazine speculates that "Cowichan knitters nowadays are more likely to buy ready-to-knit New Zealand wool from a Victoria specialty store, though some still buy their wool in long, loose sausages known as rovings and spin it themselves."

But Sylvia Olsen, who has a long history with the Cowichan knitters, explains in an interview with Cabin Boy Knits that many Cowichan knitters didn't like the New Zealand wool because it didn't have the same feel as our island wool.

Many knitters are sourcing processed roving or already-spun yarn from places like Custom Woolen Mills, MacAusland's Woollen Mills Ltd., and Briggs and Little. There is even a line of yarn produced called Prairie Sea Fusion that was developed in co-operation with Salish Fusion Knitwear, an island company creating unique Indigenous fusion handmade knitwear. Although the yarn isn't always sourced on the island, it is Canadian. The possibility of a resurgence in our own wool for these garments is appealing.

Vancouver Island and the Gulf Islands have hundreds of farms currently breeding sheep; I was able to connect with thirty-one of them raising over twenty-two different breeds. The farms I chose are all fibre conscious and interested in using their fleeces either by processing them or selling them raw, meaning they are not cleaned or processed. Some farmers have the time or manpower to skirt the fleece by picking out the dirtier parts and the larger bits of vegetable matter before selling it. Others leave that up to the maker or processor. I heard the same story many times while talking with farmers.

Because our island doesn't currently have a running fibre mill, farmers must send their fleeces to an off-island mill. And parcels of fleece are large and expensive to send. They also have to spend time prepping the fleece before they can send it so it's not full of vegetable matter and seconds. Most mills request the fibre be sorted before it's sent. The mills don't have the manpower or infrastructure to deal with the dirtiest wool. It can then take six months to a year for the farmers to get their processed fibre back. And then they have to market and sell that product while still doing all the daily farm jobs that take up their time. It is no wonder farmers tend to lean into what they know: food production.

But even the breeds that are grown specifically for protein grow a fleece, and those fleeces can often be used for something. Companies and networks are lobbying to get wool certified as insulation. Equipment such as pelletizers can turn waste wool into pelleted mulch.

With this large variety of fibre coming from just this one animal, there is no reason that Vancouver Island cannot have a thriving fibre mill. We can produce a multitude of garments for our bodies and homes, all from fibre grown on the islands. The fibre our region grows is unique to us and it deserves to be showcased by the local economy. The product is here, and the infrastructure needs to follow.

The region from which the sheep is bred and raised plays a part in the specifications of the fleece. The following information about breed-specific wool was researched using general information about the breed but also includes as much regionally specific information as possible. For example, a flock of Icelandic sheep bred and grown here on the Island will have a different quality fleece than an Icelandic sheep bred and raised in Iceland. Also, the different microclimates on our Islands and the way the animals are raised can play a part in the characteristics of the fibre.

The following resources can be used to source breed-specific fibre or fibre products. Please use these resources respectfully and only to purchase or inquire about fibre-related products.

Breed-Specific Wool

While searching for fibre, I came across many breed-specific sheep, bred over thousands of years for the unique characteristics of their fibres. The many differences include the fibre's feel, its lanolin content, its crimp (the natural wave in each individual fibre), and the way to best use it to create durable, functional, long-lasting items. While many fibre or yarn types have multiple applications, some will hold up to more wear and tear, while others are softer and more delicate, better for making finer products. Some fibres will felt and others won't. Some will take dye with lustre while others are dull, and some won't take dye at all.

Babydoll Southdown

This breed is friendly, known for their cute smiling faces, and is one of the easier ones to handle. They are a medium type of Southdown, part of the Down family of sheep. Down sheep have a relatively short staple length (average length of the individual fibres) and produce a very fine fibre. The Babydoll Southdown stands anywhere from eighteen to twenty-four inches tall, with fleece that is dense, resilient, and versatile. The one-and-a-half- to four-inch fibres can be found in white or a beautiful blend of browns. Because the fleece is dense, the locks pull apart rectangular in shape, not tapered, and tend to hold together.

▲ Babydoll ewes and lambs huddled together at FurryCreek Farm.

Opposite, top Babydoll wool in the farm store at Yellow Point Farms.

Opposite, bottom One of the picturesque barns at Yellow Point Farms. Unfortunately, this barn was lost in a tragic fire in the spring of 2023. No people or animals were harmed.

Babydolls are born either white, black, or spotted with fuzz on their heads and legs. Shiny and slick at birth, the fleece starts to fluff out within a month, and they are ready for their first shear by a year old. As the lambs grow, the black fleece lightens to a variety of browns depending on the sheep's lineage. They may stay darker if both parents were coloured, or lighter if one parent was white.

USES: The white wool takes dye nicely, not lustrous but not dull. The individual fibres have more barbs (individual ends on the scales that make up the surface of the wool), which makes this fibre great for spinning and blending. The fleece makes a fibre suitable for being worn next to the skin, such as in hats, mitts, sweaters, vests, and blankets. If you are processing your own, you can spin this fibre very fine and make a yarn suitable for infants or fine products.

Yellow Point Farms, Ladysmith

Settled in a south-facing valley is Yellow Point Farms' thirty-two ranging acres. The pastures are airy and open with naturally sloped and forested perimeters for the Babydoll sheep to graze, nurse their young, and nap. The wooden barns and outbuildings are breathtaking set against the sun and flowering fruit trees.

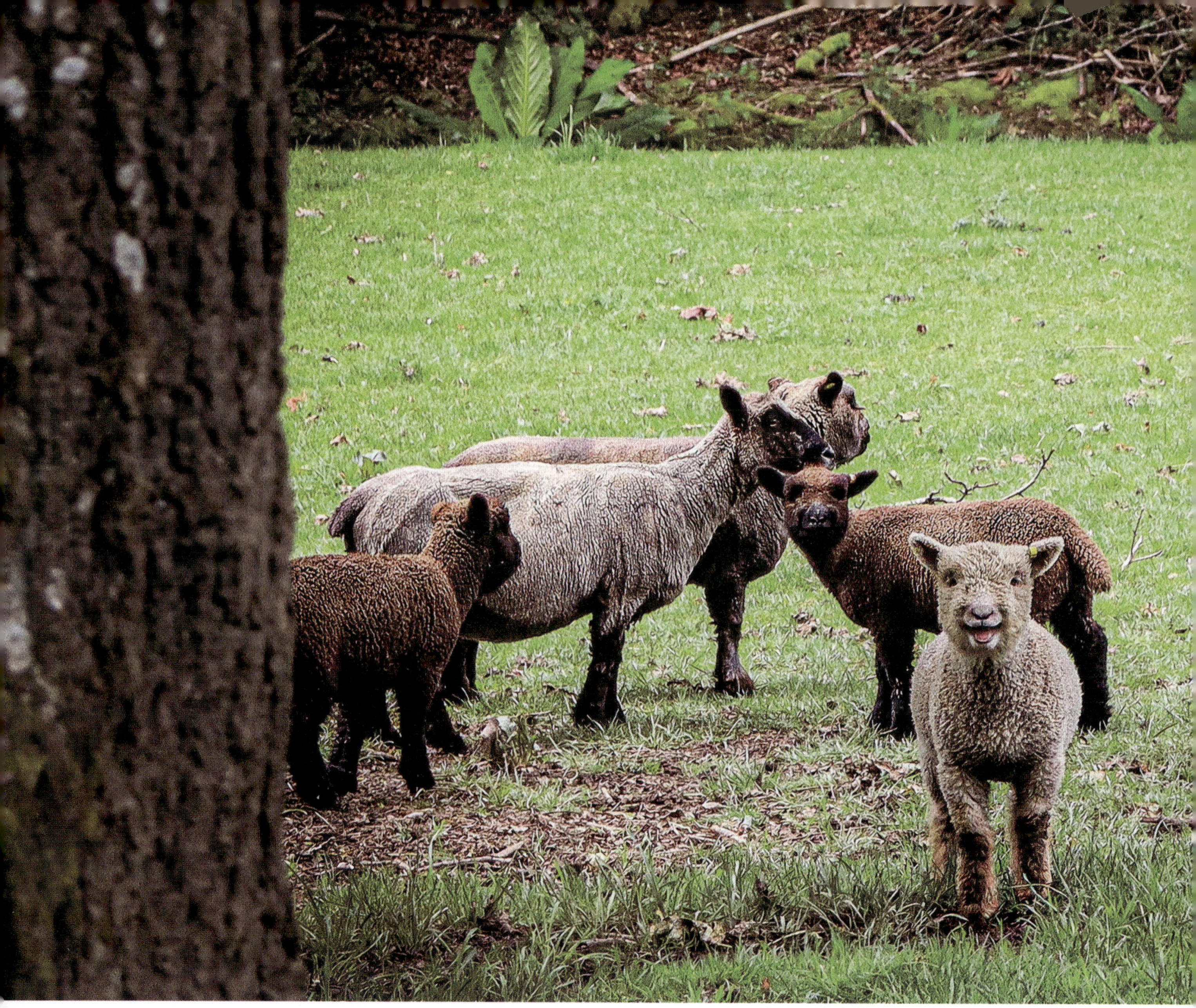

▲ A Babydoll lamb calls for her mother in the grazing pasture at FurryCreek Farm.

▲ A smiling Babydoll lamb.

Rebecca Dault takes me on a tour of the barns and back pastures. We walk out to see the grazing sheep, with lambs lying in the sun soaking up the first warmth of spring.

This family-run farm has been operated by Justin, Rebecca, and their two children since 2018. Producing wool is important to them because they strive to create a well-balanced ecosystem where all products are used in a sustainable way. Besides sheep, the farm is home to Nigerian dwarf goats, Kunekune pigs, miniature ponies, donkeys, peafowl, and a variety of poultry birds.

Yellow Point Farms has a variety of products available in their seasonal farm store, as well as at the Duncan, Cedar, and Island Roots Farmers Markets in the summer. They have a petting farm open to the public, where kids and adults can come and visit with goats, sheep, pigs, donkeys, and poultry birds.

PRODUCTS: Two-ply yarn in cream, dark, or grey, seasonal chicks, lambs, goat kids, goat milk soap, tea, garlic, blueberries, blueberry plants, eggs, honey, and other fresh, seasonal produce.

FurryCreek Farm, Ladysmith

This tranquil six-acre farm is home to a flock of twelve Southdowns—ten ewes and two rams. Don and Marva Smith take pride in their tidy property, with flowering shrubs, glowing ground covers, and a unique leaning eucalyptus. It makes the best use of the forested areas and the creek running through the centre. The sheep are free to graze in multiple pastures while staying within the protection of the surrounding trees. The farm also has a couple of goats, chickens, three Aussie shepherds, and lots of Pomeranians.

The farm focuses on breeding stock and produces the sweetest smiley lambs, sheared annually, usually in April or May. The raw fleece is available to buyers.

PRODUCTS: Raw fleece, lambs.

Black Welsh Mountain

This unique breed is the only type of Welsh Mountain sheep found in our region. The compact breed stands at about thirty inches tall. The Black Welsh Mountain grows a fleece that is generally true black, not dark brown, though it may sun bleach on the tips. Unlike some growers of other breeds who strive to eliminate the black-wool genetics from their herds, the Black Welsh Mountain is generally bred for the true black fleece. The fleece is dense and firm, yet soft and mostly kemp free (kemp being coarse, straight

▲ Black Welsh Mountain sheep relaxing by the barn at New Wave Fibre, Pender Island.

fibre). The two- to four-inch fibres have a crimp that can sometimes be hard to distinguish.

USES: The low amount of grease in this breed allows for less scouring. The fibre will create a lightweight yarn that is warm and will work for a variety of garments and household textiles. Not good for dying, as the colour is lost in the black; however, if you comb and blend the black fibre with other colours or make a two-ply with a contrasting colour, you can get some wonderful results.

Gun Barrel Canyon Alpaca Boys, Nanaimo

This ten-acre farm is run by Deb and Mark Pawlyshyn. Deb is a fifth-generation sheep farmer who also raises alpacas (see page 121). The property's long gravel driveway leads past the main home to the

▲ Mixed lamb at Gun Barrel Canyon Alpaca Boys.

shepherd's house, a small two-storey green building. On either side of the drive are pastures, barns, and treed areas for the sheep and alpacas to graze, nurse their young, and rest.

As we walk the property and pass the shepherd house, Deb explains, "It's easier to stay in the shepherd house during lambing season." With about forty-five lambs born each season, sleep can be scarce during the lambing hours, so "the closest bed is best."

The breeding focus of this flock is fibre, with Romney, Southdown, Scottish Blackface, and Valais Black Nose (F2). Deb has also bred some interesting crosses that are producing beautiful fibre. The goal of the cross is to achieve a sheep with the dense fluffy fleece of the Romney and the shine and texture of the Valais. There's also a chance of getting some mottled or brown lambs, which if they were purebread would be referred to as "spitti." Deb also wants to

SHEEP   35

▶ Handspun fibre from Gun Barrel Canyon Alpaca Boys.

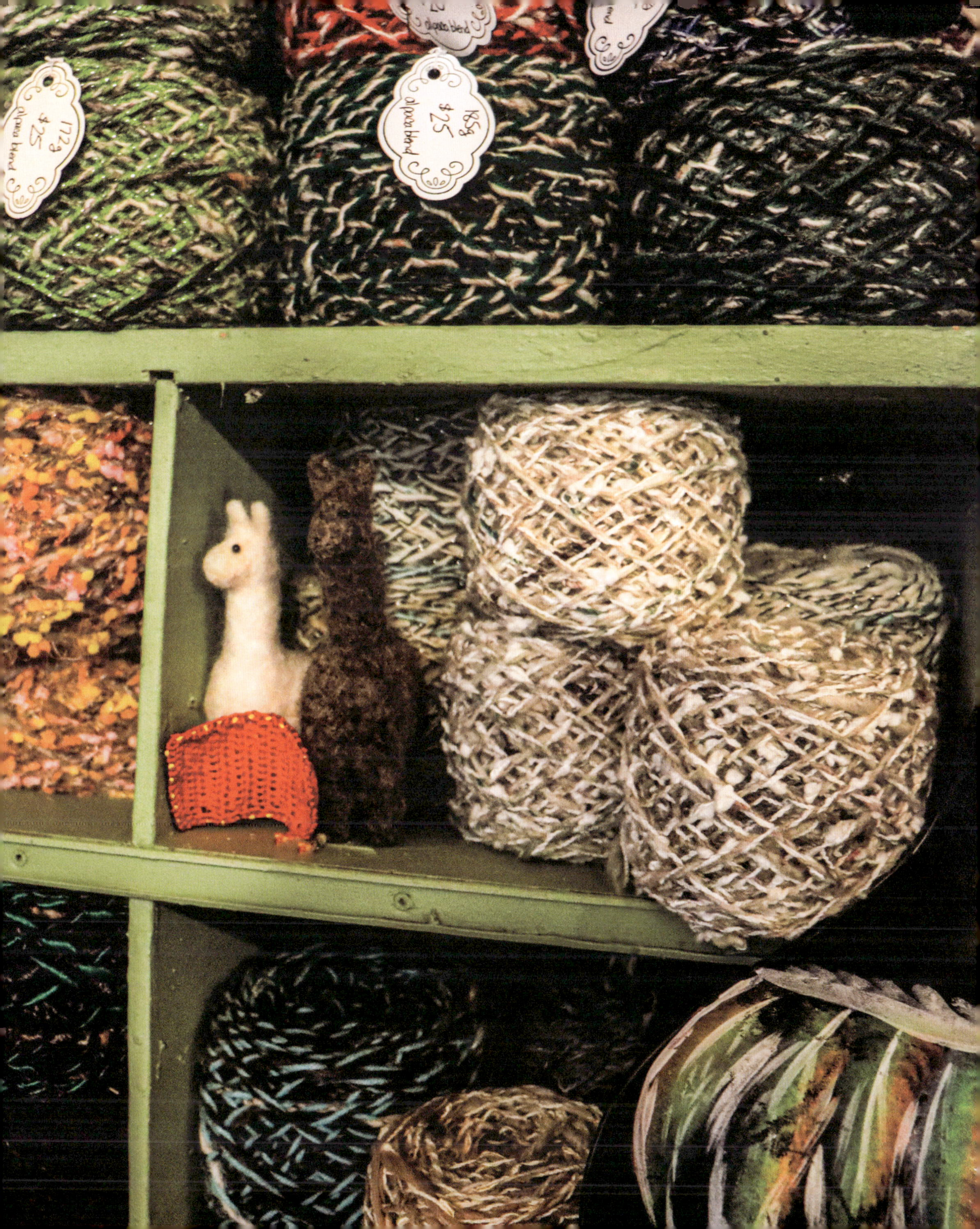
172g
$25
alpaca blend
185g
$25
alpaca blend

achieve a "polled," or hornless, flock to minimize damage to animals and property caused by sheep catching their horns on fences and brambles or ramming each other.

In the past, the farm has run several fibre-related workshops in the studio space. The farm also has a yarn and gift shop selling handmade items by hired local knitters and a variety of yarn processed on the farm.

PRODUCTS: Raw wool in season, handspun yarn from the sheep and alpacas, slippers, bed rolls, ponchos, jackets, sweaters, bum warmers, hats, mitts, and roving rugs. Deb also blends some colourful and shiny acrylics with her wool to create fun, durable yarns.

New Wave Fibre, Pender Island

New Wave Fibre includes two farms, one on Pender Island and the other on Thetis Island.

Through a winding forested driveway, I follow Jodi Schamberger to her barn. Jodi runs and works the 160-acre farm and flock on

Pender Island with thirty-plus sheep, all fibre-focused breeds. She has lived on Pender since 1996, handling, breeding, and caring for her sheep and others.

When we get to the towering wooden barn, her flock is relaxing in the shade of the building. There are three Black Welsh Mountain sheep sleeping by a rock wall as we approach, two with true black fleece and the third faded a bit with age but still very dark. Around the back of the barn we find Romeldales and a couple of California Variegated Mutants (CVMs) that are a beautiful cinnamon red and badger-faced (meaning badger-like markings around the eyes). There are some Cotswold × Romney, some Romney × Bluefaced Leicester × Cotswold. In another field we find a charming, laid-back Romeldale ram with this year's lambs. The only intact ram I am able to go in the fence with, he meanders up and says hello. Solid and quite large, with light brown, almost grey fleece and a crinkly wide nose, his demeanour is calm and non-threatening.

Jodi shows me some fleeces that are kept in the loft of the barn. Each one is unique, with a different crimp and colour variation. I can't believe how fine the fibres are. Each bag is labelled with the sheep breed and shear date. There are bags and bags of fibre ready for processing.

Jodi also shows me her two German Angora bunnies, one white, the other grey, both incredibly soft. The grey matches some of her sheep fibre and Jodi blends them into a super soft fibre. I am surprised how much crimp, or natural wave, it has.

We head out to visit Jodi's sheep that are pastured out for the summer months, and I get to see the light-grey purebred Cotswold ewe, with her Cotswold bangs and a fleece that has a much more defined curl compared to the other sheep.

◀ Cinnamon moorit and badger-faced California Variegated Mutant (CVM) ewes at New Wave Fibre, Pender Island.

PRODUCTS: Lopi, two-ply yarn, roving, German Angora rabbit fibre, ecologically sustainable fibres, organic blueberries.

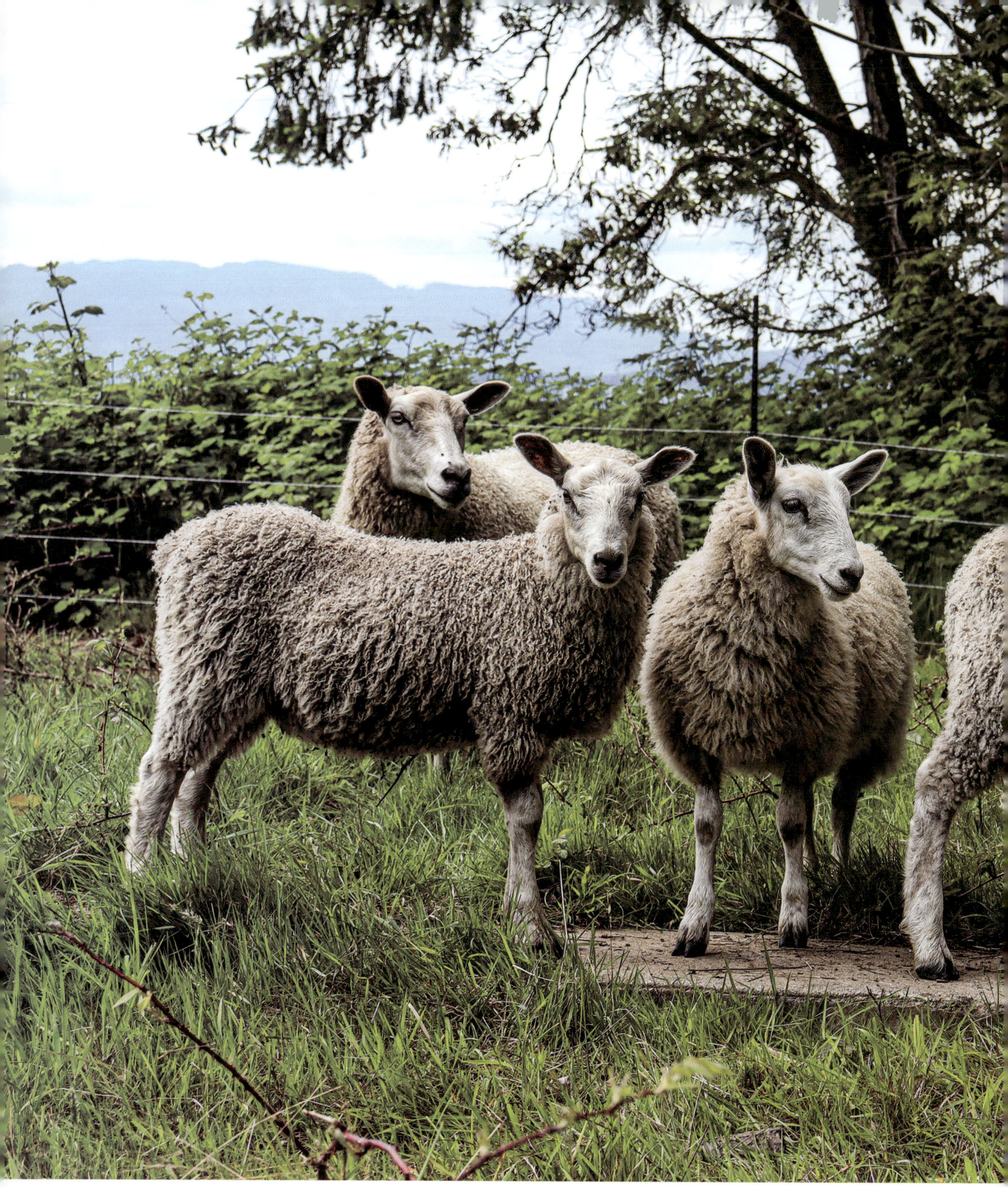

◀ Bluefaced Leicester X at New Wave Fibre, Thetis Island.

Bluefaced Leicester

This breed gets its name from its grey-blue face, which is actually its black skin showing through the short, lustrous white fibre. Its ears are large and upright, with a solid body and long back. This is one of the more predictable and versatile fibre breeds found on the island. The fleece looks like a series of tightly wound curls, fine, silky, and springy. At three to six inches, it is considered long wool. It should have little to no kemp (coarse, straight fibre) or hair in the raw fleece.

USES: Bluefaced Leicester fibre has a very lustrous shine and is known for making dye colour pop. The long staple length and crimp pattern make this fibre easy to spin. It is soft enough to be worn next to the skin and durable enough for regular use. This wool makes good clothing and household textiles, but it is not recommended for felting.

Grinsheep Fibre Productions, Errington

One of the first organic farms certified in the area, Grinsheep is run by Sharon Pickard and her family. They have lived and worked the property since the late 1990s. Beginning with a large field of you-pick raspberries and dried flowers, they transitioned to sheep and opened a fibre store.

The flock started with two foundation ewes, Southdown × Cheviot crosses that have since been bred with Romney and Bluefaced Leicester rams. Sharon tells me about the interesting dynamics she has witnessed in her flock. These two foundation ewes had personality. Right away, one was dominant, and as they bred, the dominant ewe kept coming from the same bloodline and the submissive trait stayed with the other bloodline. The flock is now around twenty-four sheep, give or take lambing season, and that hierarchy is still evident.

Sharon has cross-bred for fibre characteristics to add fleece weight, staple length, and desirable crimp. She processes all her own fleeces. When we walk out to see the lambs on the property, they wander over, curious. One of them is bottle-fed and isn't scared of humans; he comes close and checks me out.

The farm also runs a fibre arts supply store and studio, which can be found online at Grinsheep.com.

PRODUCTS: Ashford, Louet, Majacraft, and Schacht weaving tools, looms, and spinning wheels. Colourful yarn, roving, felted creations, books, workshops, and more.

◂ The upright ears, long back, and roman nose of the Bluefaced Leicester (BFL) is clear in this ewe at Elf Leatherworks.

Forage Farm, Salt Spring Island

Forage Farm is home to both Elf Leatherworks and Lorries Locks. This multifaceted farm is run by Lorrie Irwin and Andy Whitehead. Lorrie has lived on the acreage since 1979. She lived in a bus for a time while erecting a barn and later a permanent home.

The house is tucked into the land, resting under the dappled light of a towering Garry oak tree. When I pull up, Lorrie is experimenting

▲ Bluefaced Leicester ewes at Hinterland Yarn.

with a deep purple dye pot that makes my mouth water when I smell it. She explains it's the leftover juice from some concord grapes she was processing.

Lorrie leads me through a full garden with peppers hanging heavy on the plants, multiple fruit trees and a separate herb garden. She explains that their garden is their main source of food. Even the sheep get some of the homegrown vegetables.

Past the barn, we go through to an area that is fenced off into multiple grazing areas so the sheep can be moved around. Andy and Lorrie explain that this helps with grazing control, allowing them to move the animals once they have foraged the tall growth, but before they eat the lower growth.

The flock is made up of ten ewes, some pure Bluefaced Leicester and others Gotland. The Gotlands have to be sheared every six months or their fleece begins to felt. The wool is often sent to Alberta or Ontario to be processed into roving. Lorrie then hand spins and uses natural dyes to create a variety of colours.

After we check out the sheep, Lorrie and Andy show me the leather workshop, equipped with shoemaking machines, including large leather sewing machines lining the walls. They show me beautifully crafted belts, bags, and heavy aprons for metal work that Andy has designed and made. Andy has also been commissioned to do some interesting repairs on old handles and even a unique bellows with abalone eyes embedded in a carved face.

It is easy to feel at home on this farm. Lorrie and Andy are the kind of people that can make anything with their hands and enjoy doing it.

PRODUCTS: Handspun and hand-dyed skeins of Bluefaced Leicester and Gotland yarn, natural-coloured and hand-dyed roving, dryer balls, hand-knit and handwoven items, handmade leather goods.

Genesta Farm, Duncan

A large fluffy Maremma sheepdog greets me with a loud bark as I pull into the Genesta Farm driveway. He is with his flock and takes his job seriously. The owner, Dr. Helen Schwantje, is alerted that I have arrived and she greets me as I walk up to the perfectly manicured home. Helen is finishing her lunch at a large wooden table in the most amazing covered outdoor eating area. It has the feel of a hunting lodge, with a large river rock fireplace at the head of the room and antlers, horns, and skeletons neatly placed on the walls. Lanterns, cast iron cookware, tools and other exploration gear decorate the surfaces. She explains that she has never hunted an animal herself, but she had access to this large collection of antlers, horns, and skeletons through her years as the Province of British Columbia's Wildlife Veterinarian.

We chat about her current work "in retirement," which I find fascinating. She has been researching mycoplasma in wild bighorn sheep and how the bacterial species is being spread through contact with domesticated flocks. The wild sheep and goats pick up the bacteria through nose-to-nose contact with the domesticated sheep and bring it back to their wild herd, where it spreads and can have devastating effects. I can tell that Helen is passionate about her work from the way she talks about it.

We walk out to meet Helen's mixed flock of twelve ewes. The rams bred into the flock were Charollais and Ile de France and the ewes were Bluefaced Leicester. The cross gives the flock a good size and produces a substantial amount of protein and fibre. The fleeces are gorgeous, each one unique, with a well-defined crimp that is almost a curl in some. The sheep are sheared once a year around lambing, usually at the end of February or early March.

PRODUCT: Raw fleece.

Hinterland Yarn, Pender Island

A blue log home overlooks a large vegetable garden and partially forested acreage. The sunflowers must be at least ten feet tall, the squash is bulging, and an abundance of fresh vegetables is ready

▲ Yarn being coloured with Maiwa natural dyes at Hinterland Yarn.

▲ A wether at New Wave Fibre, Thetis Island.

to pick and eat. With a baby on her hip and a three-year-old boy at her heels, Hanahlie Beise walks me through her well-organized property.

We visit the small flock of purebred Bluefaced Leicesters, with three ewes and two rams. One of the rams is a new import from the US, giving our area breeders a new bloodline, which is great. All the ewes are extremely gentle with the three-year-old as he pets them and chats with them by name.

Hanahlie shows me her Watershed, Range, and Dusk yarn lines, blending undyed Canadian Rambouillet and her alpaca fibre. The colours are natural and complementary to our landscape, and the fibre is soft yet durable. I can see why so many knitters choose to design their work around this yarn. Each season, Hinterland puts out a call and knitters and fibre designers submit design proposals to work with the Hinterland yarns. Some well-loved designs have come out of the project.

We walk up to Hanahlie's backyard, which is filled with dye pots of red, blue, yellow, and brown yarn. She uses Maiwa natural dyes to produce colours that are vibrant and make me smile. The pot of indigo is dark and the yarn bathing in it looks rich and electrifying.

During my visit, Hanahlie is getting ready for the September Knit City event in Vancouver, where she will have a booth with her fibre products. Her love for fibre shows in the refined products she is producing and the care she gives her farm and animals.

PRODUCTS: Woolen-spun single-ply chunky, woolen-spun two-ply worsted, woolen-spun single-ply fingering, knit kits, patterns, duvets, pillows, and sheepskins.

New Wave Fibre, Thetis Island

When I visit Emily McIvor's farm on Thetis Island, her flock of ewes, wethers, and lambs is pastured out at a gorgeous waterfront property, where they graze alongside the saltwater shore. Her flock began with two foundation ewes that are Romney × Cotswold bred with Bluefaced Leicester to create an interesting fibre-focused cross.

The characteristics of the flock are similar to the Bluefaced Leicester breed: short hair on their legs and heads, and large, upright ears. The ewes birth one or two lambs a season. The breed is hardy and thrives during Thetis Island's wet west coast winters. The fleece looks more full and fluffy than a purebred Bluefaced Leicester, with less defined ringlets, a little more like a Romney fleece. Emily aims to shear her flock every eight to nine months, leaving the fleece with approximately a five-inch staple length.

Back at Emily's cozy wood cottage, she shows me some fleeces. On the covered front porch, she spreads a large fleece out onto a wooden sorting table, and we pick away for a moment, chatting about its qualities.

Emily explains that the way they have cross-bred with fibre in mind has resulted in a fleece with many of the desirable characteristics of each breed. It holds the Bluefaced Leicester lustre and takes dye colour true, but it is large and dense like a Romney fleece.

Emily's fibre comes in a range of natural and hand-dyed colours.

PRODUCTS: Lopi, two-ply yarn, roving, German Angora rabbit fibre, ecologically sustainable fibres.

Windsor Farms, Salt Spring Island

At the top of a rural road is a large acreage with a house at the front of the property and numerous outbuildings and barns in the back. Sheila and Darryl Windsor have owned and worked the farm since 2009. They have goats, meat and milk cows, pigs, turkeys, ducks, chickens, and of course, sheep.

Their flock is around seventeen sheep of mixed breeds, some that were surrendered to the SPCA. There is a Friesian, a Romney, some Bluefaced Leicester, and even a hair sheep whose straight shiny locks look like human hair. The fibre from the hair sheep isn't really good for anything, but the sheep is cute. They are bred mainly for dairy production.

◀ Mixed ewe from New Wave Fibre's Thetis Island flock.

▼ Handwoven blankets made by Sheila at Windsor Farms.

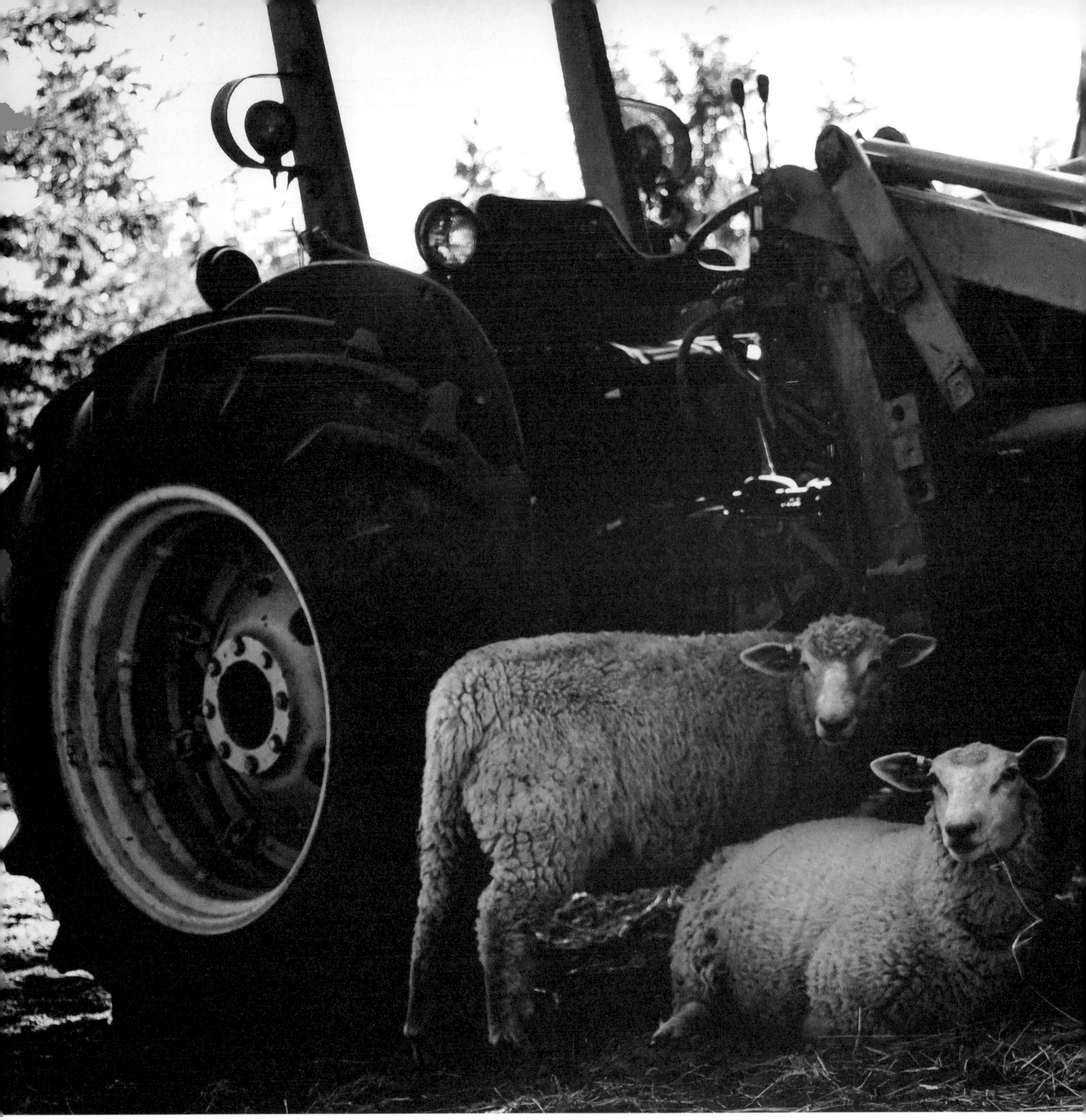

▲ Sheep at Windsor Farms relaxing under the tractor.

Sheila shows me some of her fibre work, including dryer balls, soft sheepskins, and one-of-a-kind heavy wool blankets that she weaves on her floor loom. The wool is washed and sorted before it is shipped to Custom Woolen Mills to be processed. When it comes back in two-ply white, Sheila experiments with dye colour to create some unique results.

PRODUCTS: Raw fleece, yarn, sheepskins, eggs, pork, lambs, duck meat, seasonal turkeys.

Border Cheviot

Also referred to as the South Country Cheviot, this breed is more compact than the North Country Cheviot, but they originate from the same bloodline. The breed originated in the Cheviot Hills of Scotland, where it was developed as a mountain breed suited for harsh climates and rough terrain.

The breed is fairly independent if let out to graze. The ewes make excellent mothers and need little to no help at birthing time.

USES: The fibre is good in a number of applications but is best for durable textiles. The coarser fibre can be described as "scratchy" and isn't always desirable next to the skin.

Fir Hill Farm, Pender Island

Barbara Johnstone Grimmer's Fir Hill Farm flock is made up of eighty-plus ewes, give or take depending on the time of year. She has one black Romney ewe, and the rest are either Border Cheviot or crosses. Barbara has also picked up some interesting California Red hair sheep from another farm years ago. She told me the lambs are so red they look like little Irish Setters, but as they grow, the colour turns into an oatmeal shade. Her flock can be found grazing the Pender Island golf course in the summer months.

The flock is sheared in early spring. Most of the saleable fleece is white.

PRODUCTS: Wool and wool products, grass-fed lambs, free-range chickens and eggs, seasonal fruit and flowers.

Glen Alwin Farm, Courtenay

The Glen Alwin homestead was established in 1884. The farm is a family-run business operated by three generations of women: Jo Smith, grandmother; Helen Nixon, daughter; and Clair Nixon, granddaughter. Currently 190 acres, this property is home to turkeys, cows, and a mixed flock of sheep, a cross between Border Cheviot, Arcott, and Charollais.

The farm used to send fibre away to be processed, but the cost of shipping and the undetermined time frame left them looking for alternatives. Currently, Clair is turning the farm's waste wool into a marketable, usable product right here on the island. When we spoke, she was in the process of setting up a recently purchased wool pelletizer that will turn the stockpile of wool into slow-release fertilizing pellets. Once the machine is running smoothly, Clair plans to process her backstock before offering the service to other farmers. The pellets can be found for purchase at Vancouver Island Wool Pellets.

PRODUCTS: Pelleted wool fertilizer.

Charollais

Charollais are a large breed, often with white fleece on their bodies and a thin layer of hair on their pink heads. This breed is fairly new to Canada, first arriving in 1994. Being a protein-oriented breed, the fleece is often overlooked, though it is considered fine-to-medium quality with short staples at one-and-a-half to two-and-a-half

▲ Genesta Farm's Maremma protecting his flock.

▶ The flock at Genesta Farm.

inches. When processed properly, it can create an unusually soft general-purpose yarn with good elasticity and bounce.

USES: When processing, you might consider using mini combs. The fleece will also dye true to colour but with no lustre.

Country Wools, Saanich

Lorea Tomsin got her first sheep when she was just eleven years old, and has been shearing sheep for over fifty-four years. She also shears llamas, alpacas, and goats, and is well known in the Saanich community for her work with wool. Her sheep are pastured out at as many as nine properties at a time over the summer months.

Lorea's flock is made up of pure breeds and crosses. Some of her better wool producers are the Charollais, Suffolk, Finnsheep, Friesian, and North Country Cheviot (NCC). She also has many cross breeds, including Friesian × Charollais and Cotswold × Suffolk. Although she is a wool worker and cares about the fibre, Lorea explains that fibre is "a byproduct of raising and breeding sheep" because fibre alone cannot earn back the increasing costs of hay and grain. To lower costs, Lorea usually saves up her fleeces and drives a truckload to Alberta for processing in the early fall.

Lorea can be found at the annual Saanich Fair demonstrating shearing and selling wool products.

PRODUCTS: Lopi yarn, six-strand Cowichan sweater yarn, slippers, toques, socks, hand-dyed socks, sweaters, blankets, dryer balls, and sheepskins.

Genesta Farm, Duncan
(See entry under Bluefaced Leicester, page 42.)

Glen Alwin Farm, Courtenay
(See entry under Border Cheviot, page 54.)

Corriedale

Corriedale sheep produce a medium-soft fibre with three- to six-inch staples that have mid lustre and a well-defined, even crimp that creates good loft and elasticity. The fleece is dense with undefined locks. Corriedale fleece comes in a wide variety of diameters, meaning it can range from very soft to almost crisp, most often falling in the mid-range. If you are considering this breed, it's nice to feel the fleece before purchasing so you can gauge the diameter. Fleece can be found in white, grey, and brown to nearly black.

USES: This versatile fibre works well for spinning, knitting, crocheting, and weaving. It is suitable for making clothing, socks, blankets, pillows, and household textiles.

New Wave Fibre, Pender Island
(See entry under Black Welsh Mountain, page 33.)

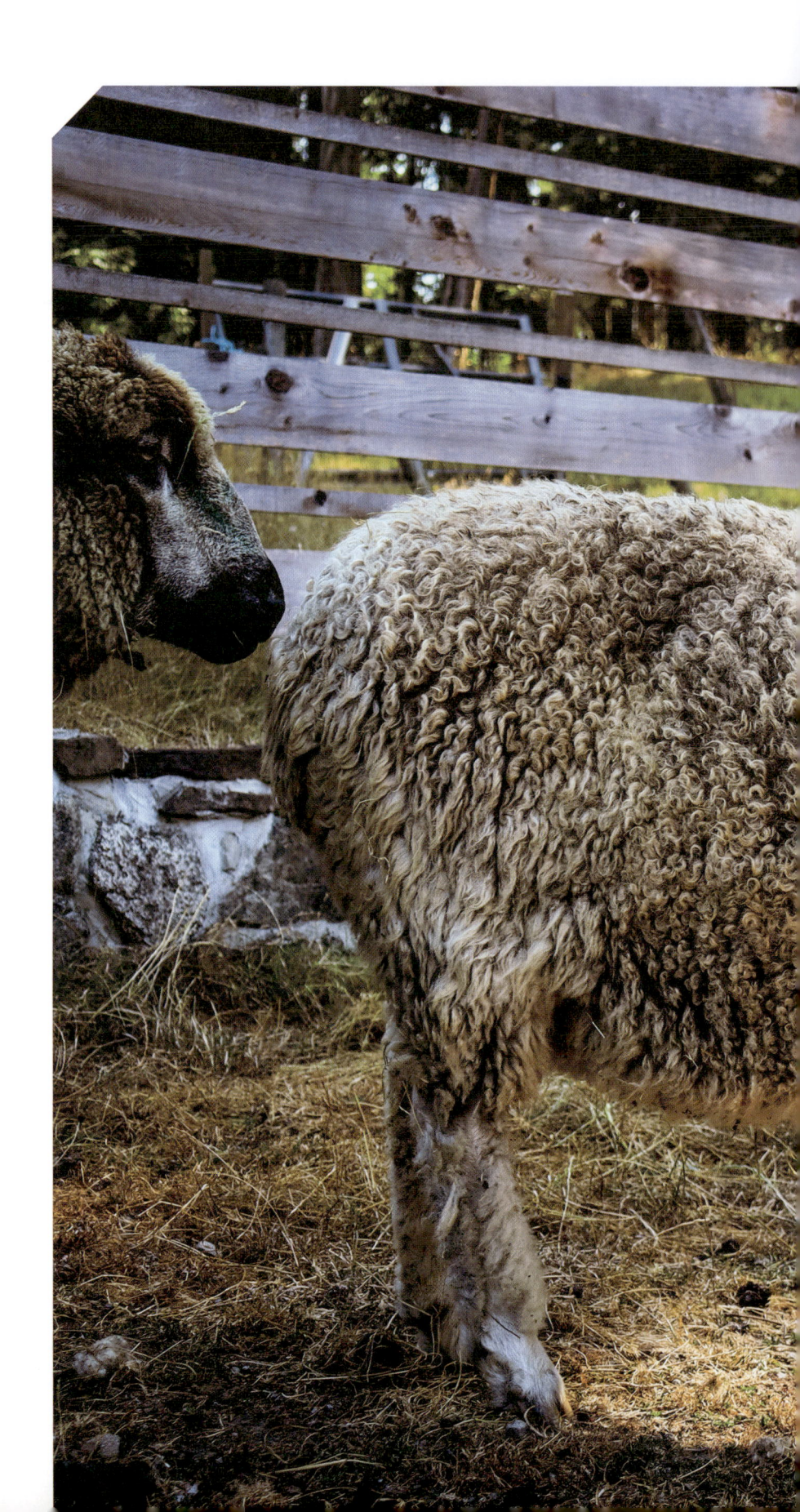

▶ Romeldale × Corriedale with some CVMS at New Wave Fibre, Pender Island.

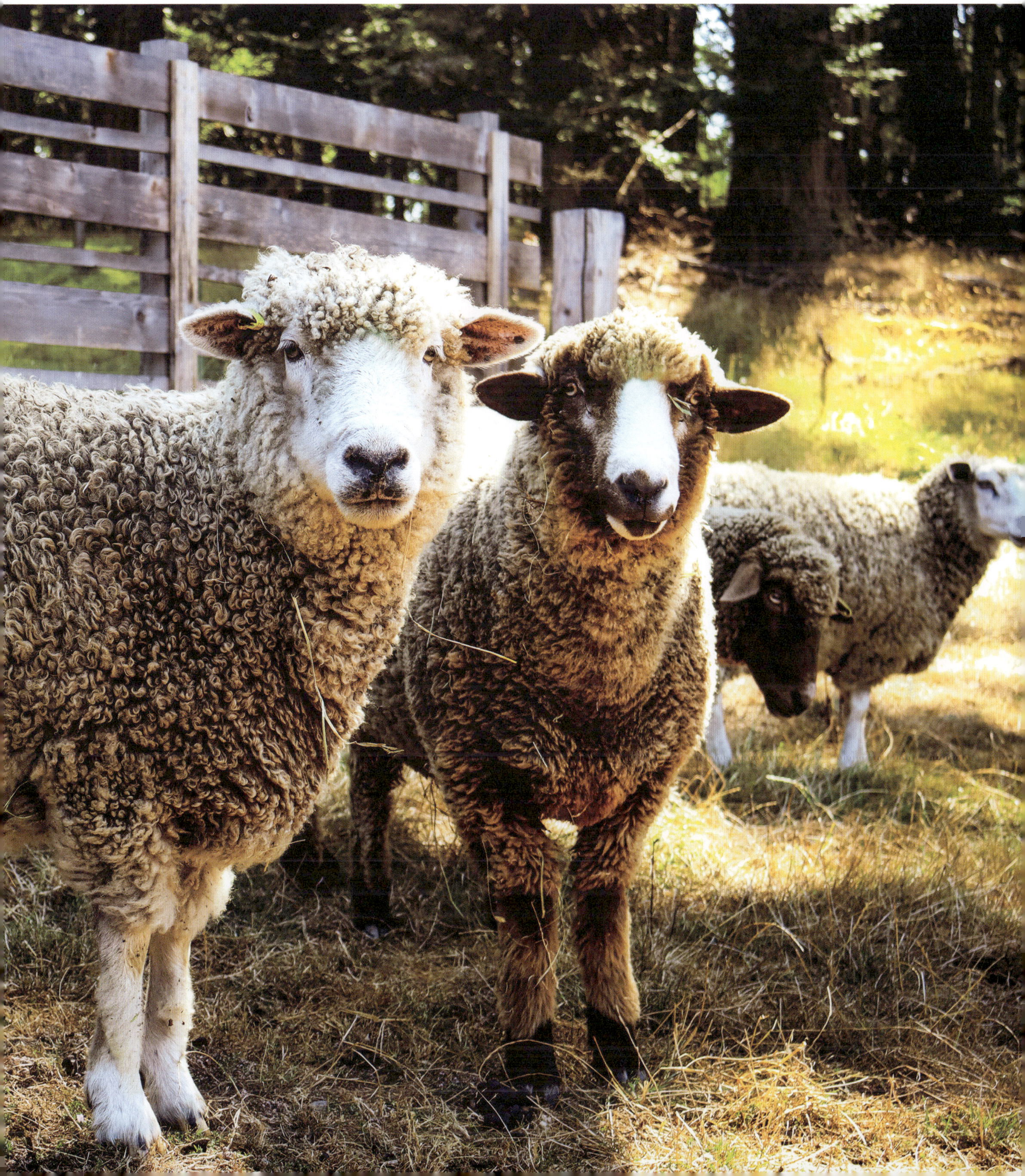

Cotswold

This breed started in the Cotswold Hills of England and was imported to North America in the nineteenth century. It was especially popular as a dual-purpose breed, for both fleece and protein, before other breeds overtook it.

  The breed is large, producing a heavy, lustrous fleece that hangs in ringlets at a staple length of seven to fifteen inches, including a full head of ringlets hanging over the sheep's eyes, giving the appearance of a funky hairdo. The length can sometimes cause the fleece to become matted and dirty, but don't let this scare you away from working with this wonderful fibre. When processed properly, the yarn has a pearl-like texture and has been referred to as more affordable mohair.

USES: This fibre can be ideal for spinning because of its length and natural curl; the fibres do not need a lot of twists to hold together. It is also great for felting, takes on colour beautifully, and is suitable for making heavyweight, hard-use items like rugs, bags, and outerwear garments. The fibre is also used in other applications where it doesn't have to be spun; for example, the locks can be woven to add texture to a project or used to create dolls' hair.

New Wave Fibre, Pender Island
(See entry under Black Welsh Mountain, page 33.)

◄ The ewe in the centre with the bangs is a purebred Cotswold.

▲ Cotswold ewe laying in the pasture.

SHEEP 61

▲ Dorset lambs laying in the pasture at Francis O Vineyard Vineyard.

Opposite, left A young Dorset ewe munching on some grapevines at Francis O Vineyard.

Opposite, right A Dorset lamb demonstrating the flehmen response.

Dorset

This medium-sized breed has been popular since it was first brought to Canada in 1860. How the breed came to be is debated. Some think it came from selected cross-breeding of Merinos and Welsh sheep, while others claim the breed was a result of a careful selection of the native southern England sheep. Either way, the sheep is a multipurpose breed that supplies protein, milk, and fibre.

Both the males and females have horns, unless polled, but the flocks I visited were mainly polled. The Dorset is one of the only breeds that gives birth multiple times a year; most breeds can only get pregnant once a year at a certain time. Dorsets are also known to give birth to twins and triplets, making them fast multipliers.

Dorset sheep have broad noses and bodies that appear muscular. They tend to produce a true white fleece with a staple length of two and a half to five inches. It has a fine crimp pattern that might feel crisp when raw. The fleece ranges in texture, making it a versatile fibre to work with.

USES: A softer fleece can make next-to-skin items like shirts, socks, hats, mitts, and blankets, while the coarser fibre is better made into outer layers like sweaters and household textiles like rugs, mats, table runners, and coasters. Because this fleece is very white it takes dye colour clearly. This fibre is flexible in preparation so you can get a variety of end results, depending on your processing method. The wool can be combed or carded, and spun worsted or woollen. Although the Dorset fibre looks like it might felt, it isn't recommended.

Francis O Vineyard, Merville

I drive up to a beautiful green barn and see Kris Welk throwing some grapevines to the rams. She greets me with a big smile and a wave. Kris and her family have lived on the farm for over twenty years. The property has been divided up and her children and their families have their own land with plenty of space to grow and live. The main house is up on a hill at the back of the property, overlooking the grazing sheep and the brilliant green grapevine orchard.

▲ A Dorset ram at Francis O Vineyard. Many of Kris's flock wear dog collars for easier handling.

Kris gives me a tour of the barn she, her husband, and her father built by hand. It's nice and cool inside and smells like fresh hay. She runs the barn's electricity on solar power and uses a gravity-fed system to water the sheep.

We walk out to see the flock and find ten ewes resting in the shade. They pay us little attention as we stand at the fence chatting.

Next, we go see the nineteen lambs in with a couple of ewes. The sheep all have long tails. Kris explains she has no need to dock them. They are laying together in the shade trying to keep the flies off their heads by snuggling. Some are munching on the grapevines Kris gave them. One of the sheep makes a funny face, curling his upper lip and exposing his teeth and gums. Kris explains that this is the flehmen response, an instinctive behaviour that allows them to take scent into a special organ at the base of their nasal cavity.

Last, we meet the rams. The intact ram is a big Dorset who comes right over to the fence and gets a nose scratch from Kris. The wether, or castrated male, is a partial hair sheep. He is smaller and I can see that his fleece is sleek.

Kris has taken up felting and is experimenting with some dye colour and wet felting.

PRODUCTS: Raw skirted fleece.

East Friesian

This breed originated in the Friesland area of Holland and Germany. It is well known for its high dairy production, which sometimes makes its wool less important to breeders. The breed is large, with no fibre on its head or legs. The fibre can come in a variety of texture grades, from fine to coarse. The three- to six-inch staple length is blocky and dense, and can be found in white or black.

USES: The white wool will take on dye colour well with a low lustre. The black is not worth dying because the colour will be lost. The nature of your particular fleece will determine the best purpose for the wool. Most Friesian is on the coarse side, making it better for durable textiles like rugs, heavy cardigans, or hard-use blankets.

Country Wools, Saanich
(See entry under Charollais, page 55.)

Swallows Keep, Galiano Island
Marcia and Ray DeVicque moved to their farm with breathtaking ocean views on Galiano Island in 2004. When they first arrived, they planted a perennial garden that now looks like it has always been there, under the towering evergreens. Overgrown ferns and

heavenly bamboo line the stone walkway to Marcia's studio gallery. She creates vivid works of glass art and one-of-a-kind jewellery. The pieces catch the light and sparkle.

In 2009, when Marcia and Ray talked about getting some sort of farm animals, they serendipitously inherited their first five sheep. The mixed flock grazes on a rock bluff overlooking the ocean. They are mixed, but mainly Romney × Icelandic Friesian. There was even a Navajo-Churro ram bred in at one point. Marcia selects her ewes for keeping by their fleece colour and staple length.

After shearing, the fleeces are stored for around three years, and then delivered—usually driven to save shipping costs—to Custom Woolen Mills to be processed. The returned wool is usually six to seven hundred skeins. The wool is left in its natural colours, ranging from three to six shades of grey to almost black, brownish, and white. The fibre is sold in the gallery.

PRODUCTS: Lopi, roving, and an Airbnb cabin.

Parry Bay Farm, Metchosin

Lorraine and John Buchanan have been raising sheep in the Greater Victoria area since the early eighties. They leased their first forty-acre property near Taylor Beach and that's where they came up with the name Parry Bay Farm. A flock that started as forty sheep quickly became more as the Buchanans found more property to lease. They now have three hundred breeding ewes that are a mixed flock of cross-bred East Friesian, Finn, Suffolk, and Texel. Lorraine explains that the East Friesian fleece tends to felt in our wet winter; cross-breeding with Suffolk adds characteristics to the fleece that make it less likely to mat up. The flock is a wide range of colours from near-black, dark brown, marbled, grey, and white.

The sheep are spread across the city and can be seen grazing in the Highlands, on the Saanich Peninsula, and in Metchosin and View Royal. They are usually moved home in December and stay there until after they lamb. When the sheep lamb in February, the flock can triple or more in size.

I follow a long driveway past open pastures, a large pond, and a white farmhouse looking over the property. Lorraine asked me to meet her in the barn, where Pieter DeMooy and his daughter Sam from Last Side Shearing are already hard at work. Pieter is set up in his shearing harness, which supports his weight as he bends over to swiftly wrestle and shear each sheep. His daughter helps guide the sheared sheep out of the barn and the next in line into the shoot. She also gathers the sheared fleeces, checking them over and sorting them into the keep or discard pile.

The fleeces to keep are thrown up to another farmhand who is standing inside a giant burlap "sausage" bag supported by a wood frame. He packs the fleeces down until the bag is full, weighing two hundred pounds, give or take. The fleeces that contain too much hair or are too dirty and matted are thrown into another bag so they don't bring down the value of the other fleeces.

Lorraine explains that they just sent thirty-seven of these burlap bags to the Canadian Co-operative Wool Growers Limited, where

◀ East Friesian X waiting to be sheared at Parry Bay Farm.

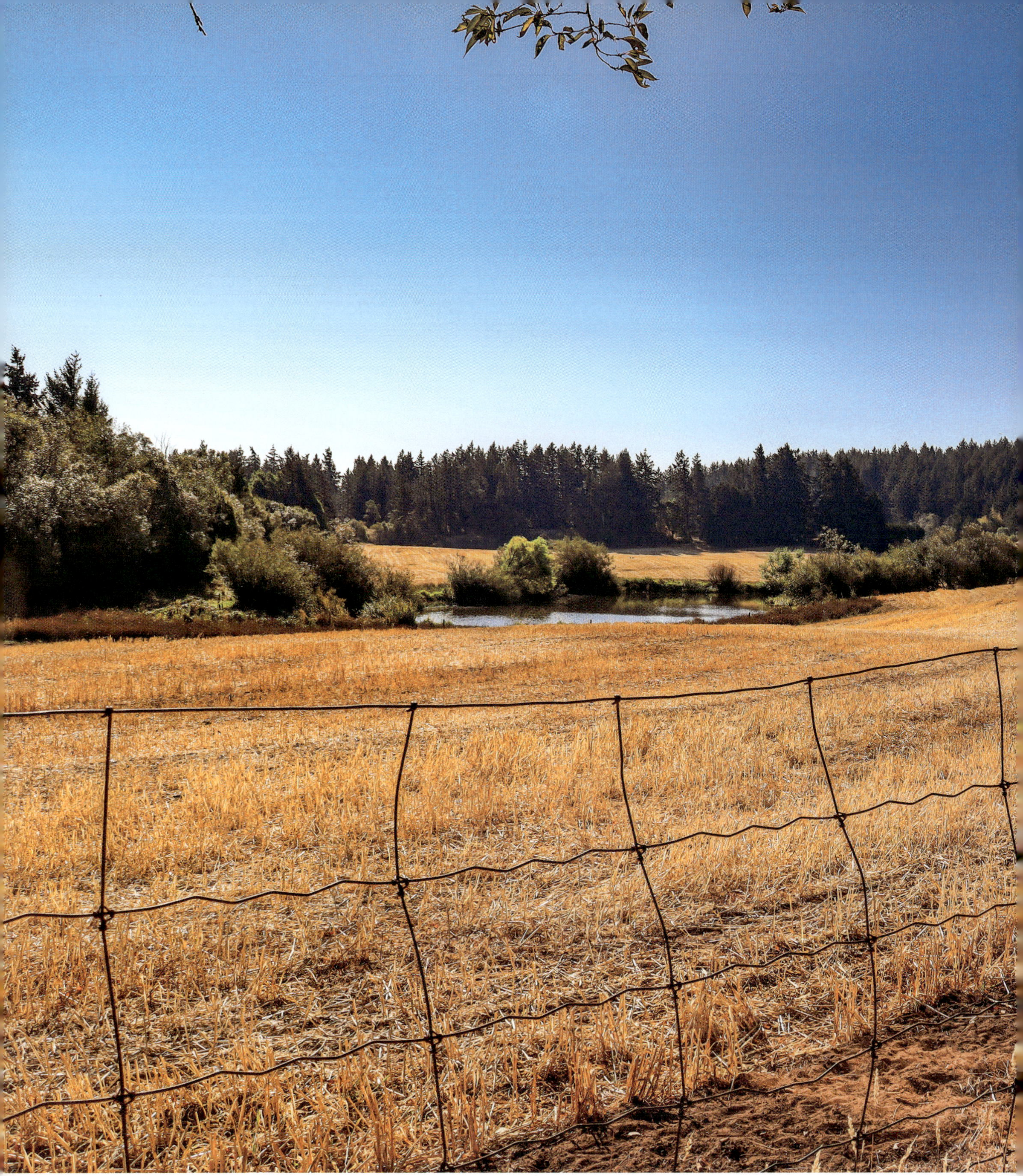

◀ Parry Bay Farm from the road.

they will be sorted, graded, and valued. The bags they are filling while I am there will be stored and sent with the next shipment.

You can find their products at their farm market, open Saturdays.

PRODUCTS: Raw wool, ethically raised meat.

Finnsheep

This Scandinavian breed first appeared in Canada in 1966. This breed is known for its prolificacy. One ewe can birth up to nine lambs, though they usually average around three or four at a time. Finnsheep can be found in a variety of colours like black, grey, and brown, but are most often white. The fleece has a staple length of three to six inches. Sometimes this breed is shorn twice a year, which results in a three- to four-inch staple length.

USES: The clean white takes dye colour wonderfully, giving the finished product a subtle sheen and defined colour. Finnsheep fibre is versatile. Soft enough to be worn next to the skin, it is a good choice for garments like tops and sweaters, or cozy blankets. If the fleece is processed properly the yarn produced can be silky soft yet durable.

◀ Finnsheep X ewe waiting to be sheared at Parry Bay Farm.

Country Wools, Saanich

(See entry under Charollais, page 55.)

Parry Bay Farm, Metchosin

(See entry under East Friesian, page 65.)

Gotland

Gotlands originated in Sweden, where the largest population still resides. They were said to come to North America in the 1920s. Farms in Canada with large flocks, like Dover Farm in Nova Scotia, were part of breed-up programs where sheep that were a percentage Gotland were imported and the crosses bred out over the years, leaving purebred Gotlands.

This medium-sized breed is curious and friendly. Its locks can be found in a variety of lengths ranging from three to seven inches, depending on shearing frequency. This breed is often shorn twice a year to produce a more consistent staple length of three to four inches. The colour of Gotlands is often in the grey family, from silvery grey to charcoal or near-black. They can also be found in white or brown, and the fleece does not sun bleach. Gotlands have a unique crimp and texture with a curly appearance that can resemble fine mohair.

USES: Suitable for creating next-to-the-skin garments.

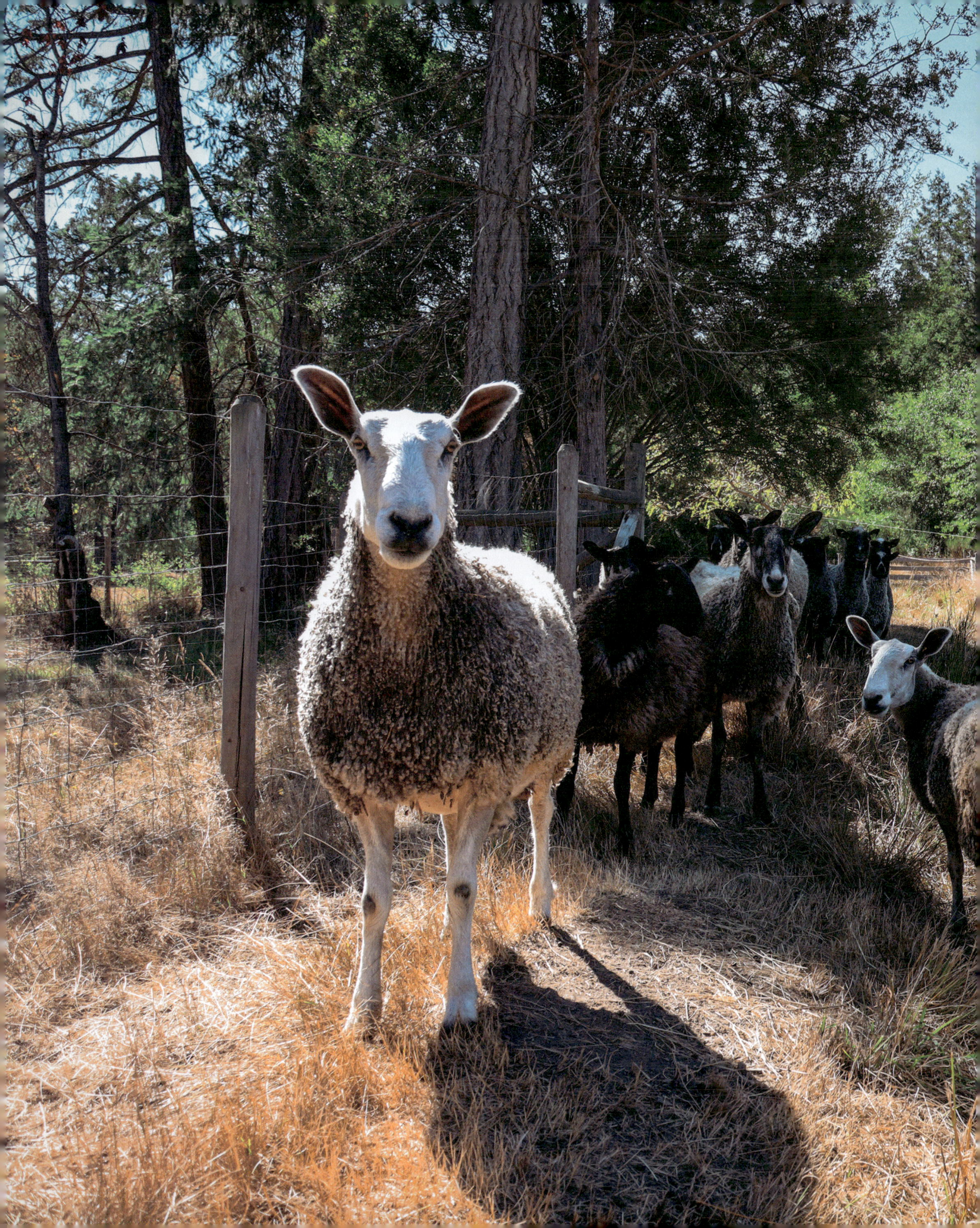

◀ This flock of Bluefaced Leicester and Gotland ewes are strategically grazing in their designated pasture at Elf Leatherworks.

Forage Farm, Salt Spring Island
(See entry under Bluefaced Leicester, page 42.)

Tsolum River Ranch, Courtenay
This newly established, twenty-six-acre ranch is owned and operated by Kelsey Epp. Kelsey grew up ranching with her family and knew it was something she would continue doing throughout her life. She also leases land in the area and has been working with local farms and livestock in the Comox Valley for years.

As I enter the property, a large ram catches my eye. His size and good looks make him noticeable, and I will learn that he is new to the flock. He is in with a smaller ram from the previous lambing season that will be kept with the flock. Both rams will be used to increase the flock size in the upcoming lambing season, the larger one being the guaranteed producer. Kelsey explains that breeding usually happens in October or November and lambing in March or April.

We walk out to the back pastures to see the ewes. "Woolly, woolly, woolly," Kelsey calls, and the sheep all come running. Though she has about fifty sheep now, Kelsey would like to increase her flock size, and she has the space for it. The back pastures sprawl out past the wooded area at the front of the property and slope down gently to an area that gets marsh-like in the winter.

The flock has a small number of Gotlands that are a gorgeous mix of colours. Kelsey shows me their recently shorn fleeces; they are so lustrous they glow.

◀ Gotland ewes at Elf Leatherworks farm.

▲ Gotland fleece from Tsolum River Ranch flock.

▶ Icelandic ewe at Old Road Farm.

Kelsey tells me about some interesting work she and a veterinarian friend have been experimenting with on her property. They have been testing and observing how rotational grazing and stockpile foraging can help with decreasing the number of parasites, like barber's pole worms, in the flock. By grazing only 60 percent of an area and then moving the animals, parasites have a harder time thriving, and in the long run fewer are ingested. Kelsey hasn't had to deworm her flock as much with this method.

When the sheep are shorn, Kelsey sends the fibre away to Custom Woolen Mills to be turned into roving and batts. She is also a talented knitter, felter, and designer. Her work can be found at Aase Studio.

PRODUCTS: Raw Gotland, Suffolk fleece, roving, batts, handmade wool slippers, hand-knit mohair sweaters, horsehair earrings, wool duvets, dryer balls, wool scrubbies, softspun yarn, handspun yarn.

Hampshire

This breed was developed around 1829 in Hampshire County in England. The breed came from crossing the Old Hampshires with the Southdown and then again with Cotswold. The resulting breed is large and solid with a blocky build, most often found with white fleece and characteristically black noses, ears, and legs. Hampshire fleece is dense and medium-grade, with blocky, rectangular-shaped staples that average two to four inches long.

USES: This wool will take dye well, being neither lustrous nor flat. The yarn can be used to make socks, mitts, hats, sweaters, and everyday items that require durability. This fleece can often be overlooked by spinners, but it can be fun to process, and it gives a versatile result.

Gun Barrel Canyon Alpaca Boys, Nanaimo
(See entry under Black Welsh Mountain, page 33.)

Icelandic

This well-known fibre breed with its unique double-coated fleece was brought to Canada in 1985. The top outer layer of the fleece is called *tog*; strong and water-resistant, it can be silky or mohair-like, and wavy with little to no crimp, more like hair. The finer undercoat is called *pel* and is super soft, more like fine wool. When blended, these two types of fibre from one double fleece make *lopi* yarn.

This breed is sheared twice a year, once in the spring and again in the late summer or early fall. The summer fleece tends to be cleaner and easier to process, as it tends to get a bit matted and muddy during our wet winters. The staple length for this breed varies and the fibre can come in a range of colours. White is predominant but black, brown, and badger are also common on the islands.

USES: Tog is good for making durable hard-wearing items. Pel is good for making baby garments, underwear, soft shawls, and lace. Lopi is durable, soft wool that is the staple wool of traditional Icelandic sweaters and garments.

Mountain View Icelandics, Merville

A large, low-standing home sits at the front of the Mountain View property with a green barn by its side. The buildings are tucked into the greenery of tall cedars and well-established gardens. I can see towering sunflowers in the full veggie garden.

Anne Everett greets me at the front gate, and I follow her to the barn. She has a fleece with gorgeous colour variations out on a sorting table. She shows me samples of the other fleeces from her flock, holding them up to the light. Each one has its own unique colour, length, and texture. Anne is a spinner and tells me she has gotten tweed-like results from spinning some of the fibre without blending it.

We take the gator (UTV) out to see the sheep in the back of the property. The trees open up past the barn and we are in twenty acres of open, airy pastures. We go out and sit in the grass with the eleven ewes, who are alert and curious about us. They aren't big sheep. Anne lures them over with some grain from a pail. Hers is a fibre flock, and she tells me how she is buying lambs for colour. They are sheared in September and April, and have just been shorn, but I can tell that their fleeces are all unique. Some lie down in the grass by us and others keep grazing. We watch them for a while, enjoying the sound of them chewing their mouthfuls of grass and grain, before heading back to the house.

Anne's farm is also home to her business, Heads Up Kennels, which has been breeding and training competitive and working golden retrievers for over twenty years.

PRODUCTS: Raw Icelandic fleece.

Old Road Farm, Duncan

This twenty-two-acre farm has been home to sheep for decades. Walking past the original white farmhouse and into the back of the property, we pass outbuildings covered in overgrown Virginia creeper that I can only imagine in full bloom.

Opposite, top Icelandic fleece on the sorting table at Mountain View Icelandics.

Opposite, middle A lock of Icelandic fleece. You can see the tog and pel holding together.

Opposite, bottom Ultreïa Farms Icelandic lopi in four natural colours

▲ Icelandic ram in the back pasture at Rabbit Hole Farm.

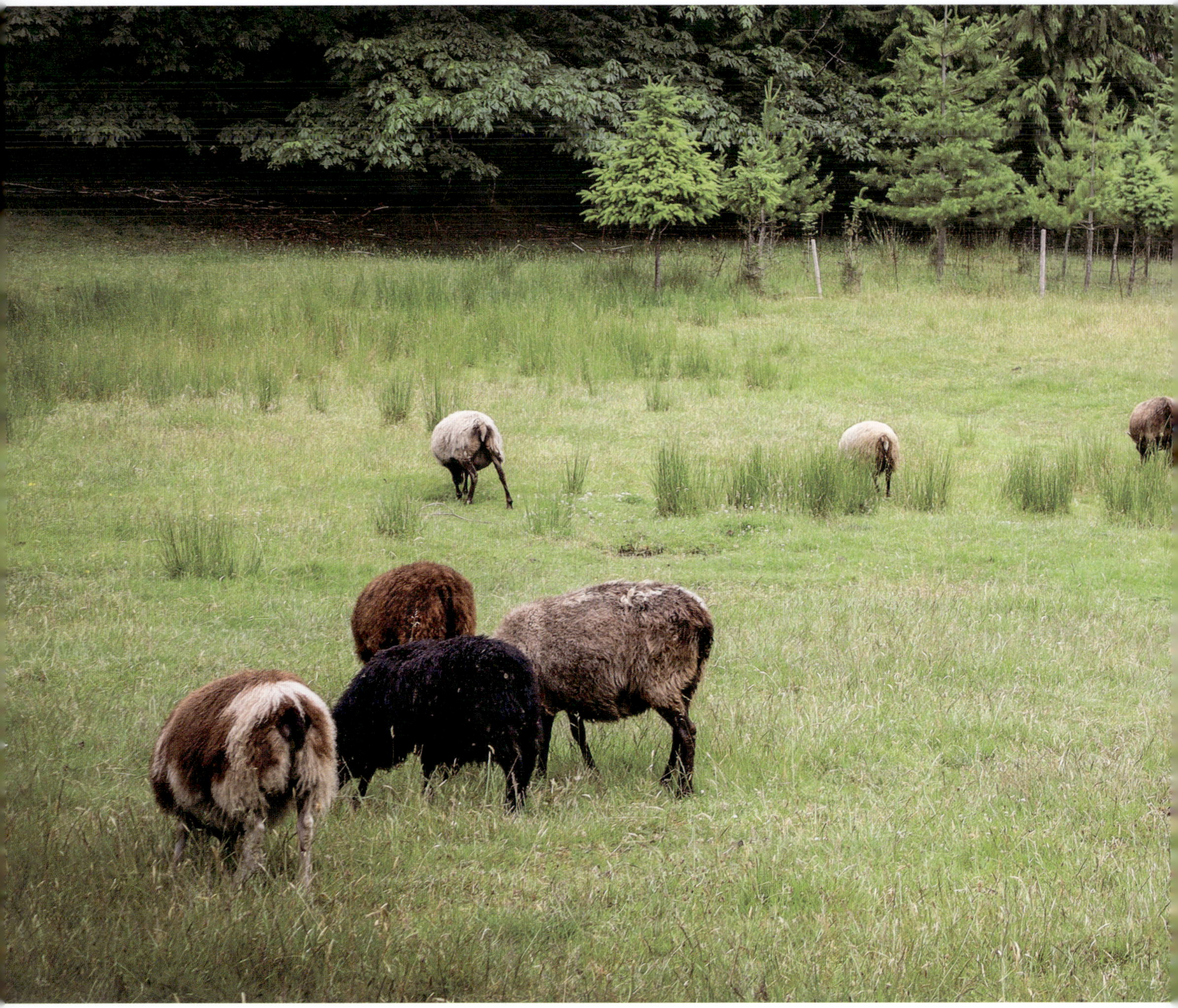

▲ Icelandic sheep grazing at Old Road Farm.

▲ Freshly shorn Icelandic ewe nursing her lamb at Ultreïa Farms.

◀ Young Icelandic ewe at Rabbit Hole Farm.

▶ Icelandic ram in the back pasture at Rabbit Hole Farm.

Owners Harry Williams and Joan Kallis have converted an old barn into a ranch-style home that overlooks the back pastures and ponds. Harry is a second-generation sheep farmer who grew up on the property. For a while, life took him away from the farm, but when it was time to settle and raise a family, he found himself back in the Cowichan Valley.

The farm is home to seven Icelandic ewes and their lambs. The colours in this flock are gorgeous—spotted, black, brown, caramel, and grey. The sheep are sheared twice a year and their fleece grows exceptionally fast, giving the staple close to a six-inch average.

PRODUCT: Raw fleece.

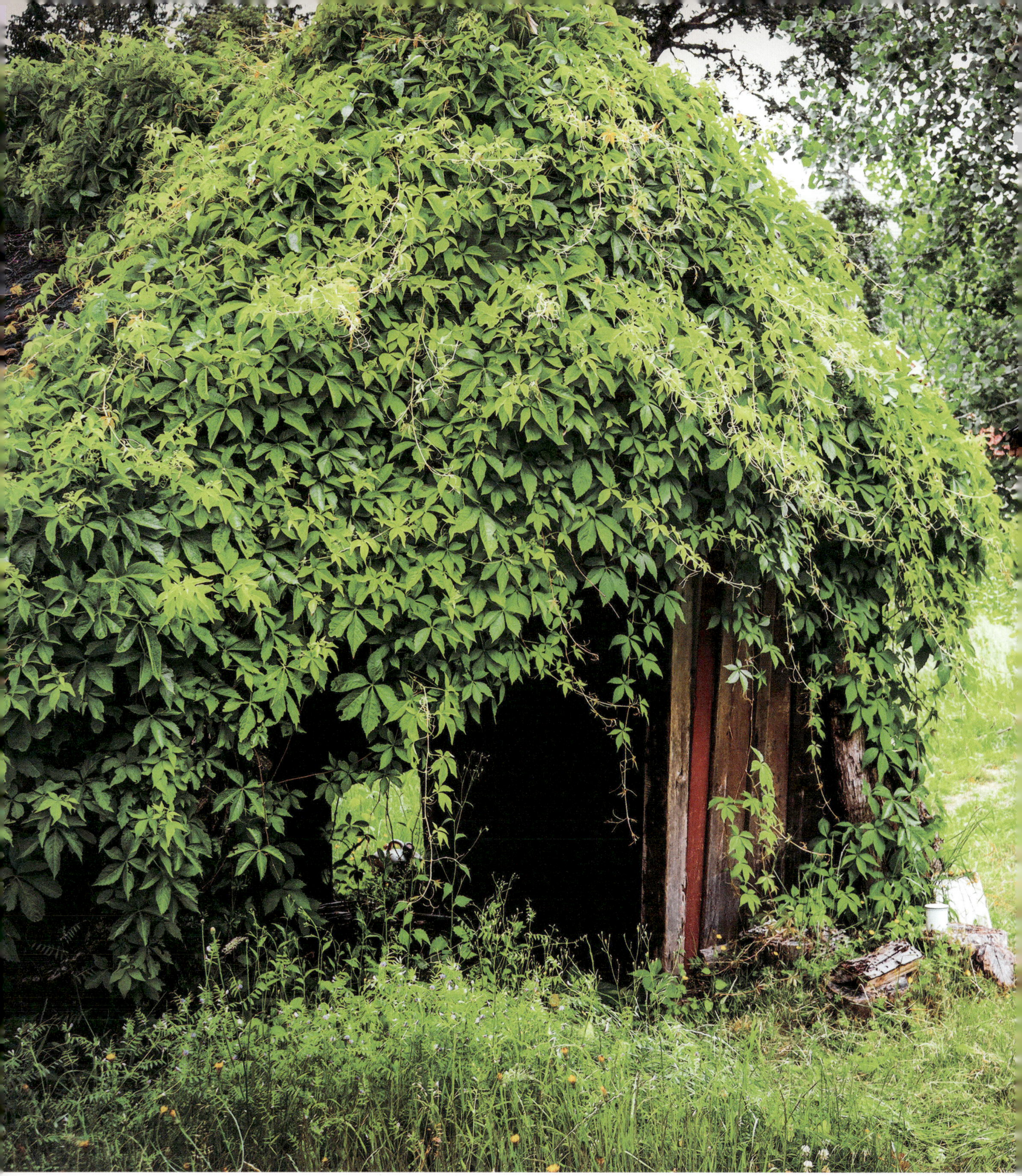

◀ Virginia Creeper taking over an old building at Old Road Farm.

▶ Rooster at Rabbit Hole Farm.

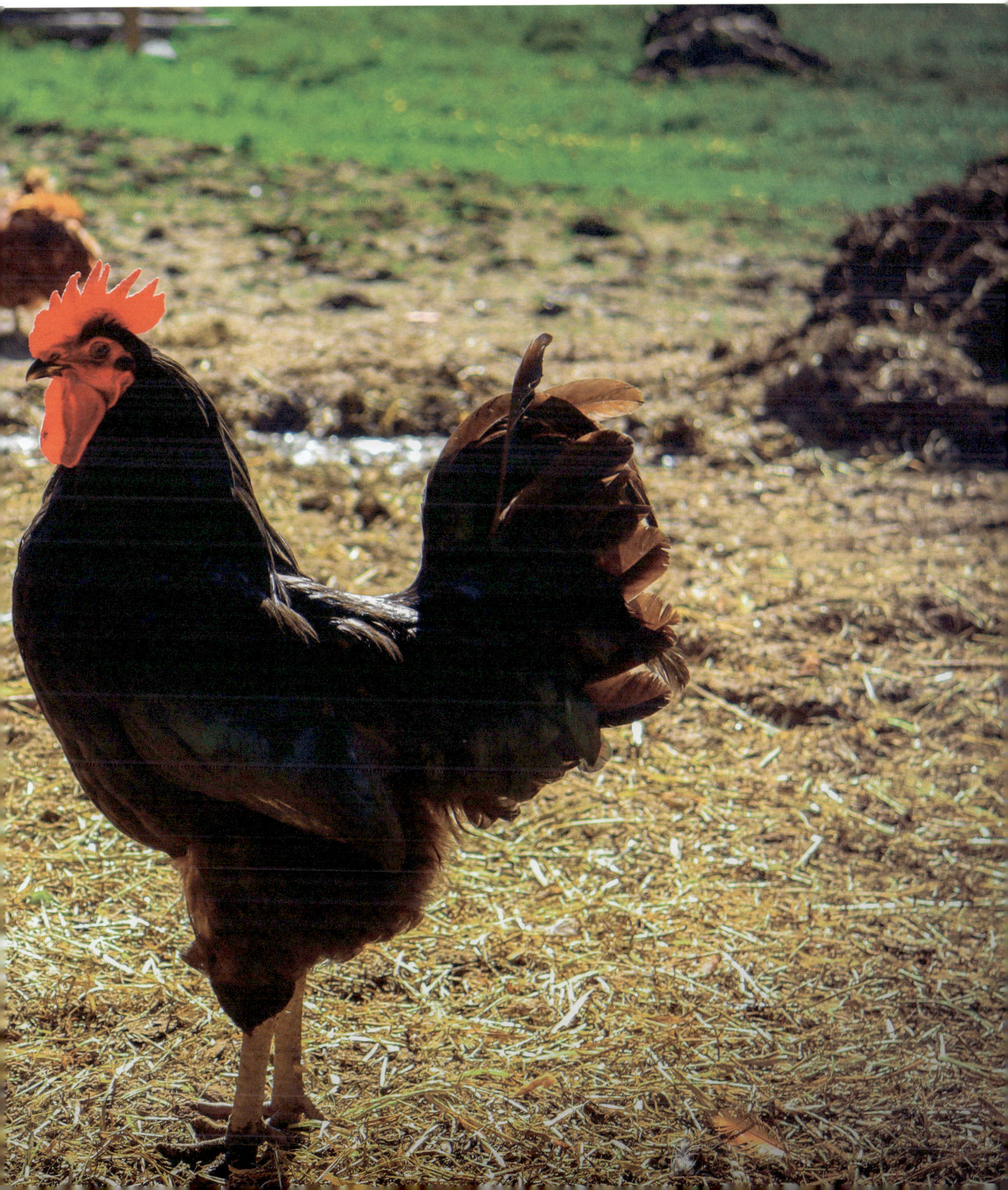

Rabbit Hole Farm, Cowichan Bay

Becky and Jonathan Lomas moved their family back to Vancouver Island in 2021 and set down roots in the Cowichan Valley. Their eight-acre farm has an airy, spacious feel, with pastures set behind the main house and barn, and a yard filled with the sounds of geese, chickens, sheep, pigs, and cows. The animals are used to the family being around and the kids can pick up the lambs and pet the ewes.

The two rams roam the back pasture with three mini-Hereford cows. They are friendly and curious when we approach the fence. Becky is worried she may have to put one of the rams down. His horns are curling at a tight angle, pushing into his face, and when the two rams butt heads, the horns dig into his cheeks and do damage. He seems quite happy at the moment, but it's a worry.

The ewes and lambs have the run of the barn with the chickens. They happily eat the hay and curl up in the straw bedding. The sheep are sheared twice annually, winter and summer. The summer coat is always a bit cleaner because of the weather, but the winter coats produce a large amount of usable fibre and shouldn't be overlooked.

PRODUCTS: Icelandic winter and summer fleece, purebred register Icelandic ewe lambs, Icelandic ram lambs, mini-Hereford breeding stock, piglets, goose ganders, and chicken eggs.

Swallows Keep, Galiano

(See entry under East Friesian, page 65.)

Ultreïa Icelandics & Heritage Poultry, Chemainus

This bustling farm has everything a herd of Icelandic sheep could ask for. Past the flowering apple trees, garden beds full of strawberries, and newly sprouted vegetables, we walk back to where the animals are housed. The L-shaped property is cut through by a flowing stream, lined with overtaking yellow flag irises, that comes down from Fuller Lake. The back of the property is forested and private, giving the sheep plenty of space to graze and relax in the shady trees.

Owners Michelle and Derek Masselink have been running the Chemainus farm since 2017, after farming on Pender Island. They have always endeavoured to process as much of their fibre as possible. They send their fleece to Custom Woolen Mills to be processed into a lopi yarn. The fleeces that are too dirty or damaged to be processed are incorporated into the farm's compost, which then gets used in the vegetable gardens.

The farm is also home to horses, mini ponies, and heritage poultry.

PRODUCTS: Icelandic wool and fleece, lambs, eggs, fresh produce.

North Country Cheviot

This breed came to Canada in 1944, when MacDonald College in Quebec imported ten ewes and two rams. Not long after, Canada's Department of Agriculture imported several of this breed, which quickly became popular with Canadian shepherds. The breed is hardy, long-lived, and long-producing, often still breeding and producing usable fleece into their old age. Though the Cheviot is primarily grown for its meat, the breed produces good-quality wool with a staple length of three-and-a-half to six inches. The staple is rectangular but tapers slightly at the tips.

USES: The fibre is versatile and easy to work with for spinners. It has been described as chalky but it will still take up dye colour with no lustre. Yarns made from this wool make good everyday items like socks, sweaters, blankets, and pillows. Though yarn from the coarse fleeces might not be suitable for wearing in direct contact with the skin, it works for outer layers and durable items. This is a good felting fleece.

Campbell Farm, Saturna Island

Scenic Campbell Farm sprawls over five hundred acres, with ocean views and hovering mountain ranges. Jacques (prounounced "Jackie") Campbell and her family live on and operate the farm, the second generation to work the land. Her parents started on the farm in 1958, and built an abattoir that processes cattle, sheep, lamb, and goats.

The flock has around one hundred Suffolk × Cheviot ewes that are bred each season and sheared in May, producing a mountain of fibre. The fleeces are white, brown, and greyish.

Family and friends have helped Jacques and her sister Nan Logan sort and prepare the fibre for shipping to various locations, depending on what they are having made.

PRODUCTS: Wool batts, yarn, hand-dyed wool socks, wool blankets.

Conheath Farm, Duncan

Perched on a sunny slope in the Kinsol Valley, this acreage is home to a small flock of North Country Cheviots. Heather Hanning chose to breed and raise them because they are a hardy breed suitable for this terrain. They also make excellent mothers, needing little to no help at birthing time.

Heather has been raising sheep since 2001 and processes the fibre on her property. She has an electric drum tumbler for removing the initial dirtier bits and vegetable matter, a flicker that removes more bits and opens the locks, and a carder she uses to make fluffy, dreamy roving—cleaner than any home-processed roving I have ever seen.

Some of the fibre is run through a felting machine, which turns it into flat, even sheets of felt that can be used to felt an image onto or sewn into other felted items.

Opposite, top North Country Cheviot ewes coming down from the hill at Conheath Farm.

Opposite, bottom Dye colour samples at Conheath Farm.

▲ North Country Cheviot at Conheath Farm.

▶ North Country Cheviot sheep grazing at Coneygeers Farm.

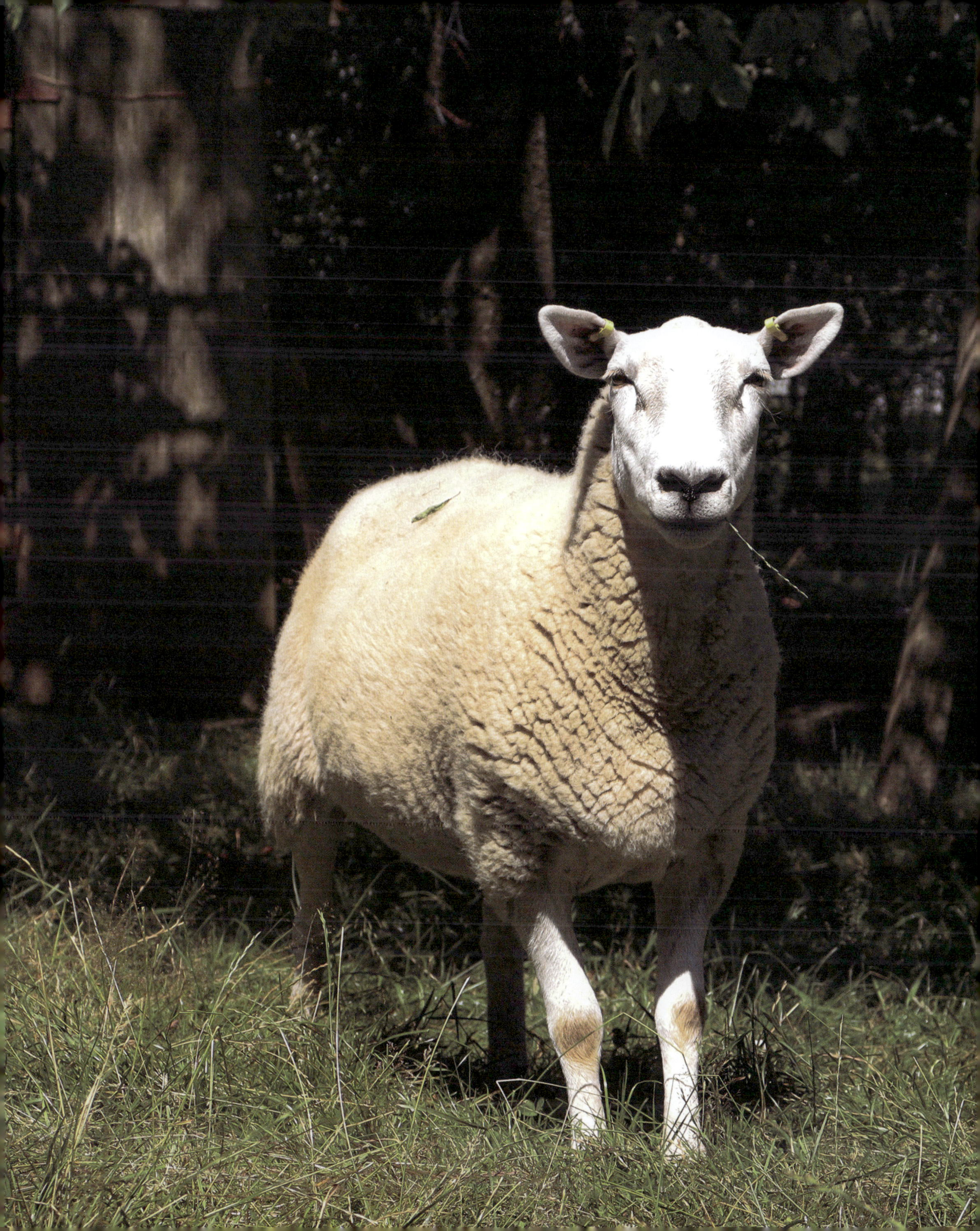

▲ Mixed breed flock in the barn at Ruckle Heritage Farm.

◀ A young highland cow at Ruckle Heritage Farm.

Heather has been working with natural and commercial dyes for years, making over fifty different colours from things like coffee, onions, cochineal, usnea, indigo, and marigolds.

She has a large selection of coloured roving, which she spins into a durable yarn that can be used to make hard-wearing items. Heather says it makes for excellent dry felting material, but it takes a long time to wet felt.

PRODUCTS: Coloured roving, sheets of felted material, and felting and dying workshops.

Coneygeers Farm, Nanaimo

The towering apple trees and fig bushes that line the outer walls of the property make it feel private and like a country farm. A small blue barn is nestled behind the white Tudor house. A true English garden overtakes the front yard, roses climbing the walls of the house.

Coneygeers was a parcel of a larger lot known as Meredith Farm before the land was split and the roads around it were built in the early 1990s. Deborah and Don Wytinck have lived on the property since the 1980s. The name Coneygeers translates to "meadow where the rabbits play." It fits the farm because their North Country Cheviot lambs, with their big ears, look like rabbits in the field, and for a time they raised French Angora rabbits.

The farm's focus was breeding North Country Cheviots for show, and Don has a wall covered in awards for their high breeding standards. At one time, they imported semen from some of the strongest British bloodlines to strengthen their flock. Deborah served on the board of the Inter Island Sheep Breeders Association as well as other sheep breeding associations. Between the two of them, they have endless stories to tell about sheep shows, their years of working on the farm, and how and to whom they've sold their fibre. As the couple ages they have moved away from showing their sheep but continue to breed a small number each season.

We sit in the dappled light of a large tree and Deborah explains to me the different parts of the fleece. The North Country Cheviot is a strong fibre, good for making outerwear items, but the fibre near the neck of the fleece is the finest and can be sorted and used to make finer wool. She also explained that the harsher the weather and winter, the more coarse the fleece.

Coneygeers sheep are generally sheared in May before lambing. The fleece has an average staple of three to six inches, with a well-defined crimp.

PRODUCTS: Raw wool.

Islandia Farm, Gabriola Island

This nearly two-hundred-acre farm is a designated Century Farm, one of the oldest in our region, raising sheep since the early 1850s. When the farm was first established and prospering, it was one of the main food suppliers to the Nanaimo coal mine industry, shipping fresh meat and vegetables to feed the workers.

In 1994, Russell Hollingsworth and Arabella Campbell purchased a portion of the property with the intention of restoring it and increasing the food, livestock, and hay productivity. Recently, they were able to acquire the remaining portion of the land from the original family and are continuing to re-establish the homestead.

To access more of the property, Russell and Arabella have used an ancient engineering technique used by the Romans and revived recently in Scotland. Waste wool is laid down as the initial foundation for roads and walkways, acting as a barrier between the wet, soggy land and the stone surface laid overtop. The wool is an eco-friendly replacement for the common plastic membrane, allowing moisture to pass through without washing away roads.

The farm currently houses a flock of sixty North Country Cheviot × Suffolk. Sheared in June, the flock is predominantly white with some brown-black in the mix.

Both Arabella and Russell are passionate about traditional crafts and small-scale circular production. They share with me that "at Islandia Farm, we see natural wool fibre as a wonder of nature and a potential panacea to the damage caused by the micro-fibre industry."

◄ One of the many pastures and buildings of Ruckle Heritage Farm.

PRODUCTS: Raw wool.

Ruckle Heritage Farm, Salt Spring Island

This twelve-hundred-acre farmstead was established in 1872 by Gordon Ruckle, one of Salt Spring Island's first settlers. The property has an interesting history, with many original buildings and homes still standing. Visitors are welcome to walk around the trails and buildings, check out the little museum, and camp near the beach. There are also overnight accommodations in some of the old heritage houses.

In 1974, the land was left to the British Columbia provincial park service with the stipulation that a parcel of land continues to be actively farmed. The now 202-acre farm is run and operated by Mike and Marjorie Lane. Mike has lived and worked on the property for over thirty-two years, and was entrusted with its care by the remaining Ruckles after the final family member passed in 2018.

The farm is home to turkeys, Highland and Jersey cows, and goats. The large flock of North Country Cheviot, Suffolk, and Texel crosses includes around ninety adults and forty lambs, depending on the time of year.

As Mike and I walk past the golden pastures and down a dirt road to let the sheep out of the barn, we talk about regenerative practices and how the animals play an integral role in the cycle of the farm. They eat the tender greens with the morning dew and then fertilize and add moisture back to the land with their waste.

The farm isn't necessarily fibre focused, but at one time it was home to over seventy black Corriedales whose fleece was sent to the Modeste Indian Sweater & Crafts Ltd. in Cowichan for sweaters. They now save the best lamb fleece from each season to sell to spinners and weavers.

PRODUCTS: Raw wool, seasonal fruit and vegetables, herbs, seeds, lavender, catnip, and fresh and dried flowers.

Romeldale

This breed is an interesting mix of Rambouillet and Romney. Medium to large in build, they have fluffy fibre on the tops of their heads and on the sides of their faces, with a bare, broad nose. The fleece is dense but soft, with long, blocky locks that are sometimes tapered, producing a uniform staple length of three to six inches. The crimp (waviness) is reliable from base to tip and there should be no hair or kemp (coarse, straight fibre). Romeldales are often white, with a coloured or patterned version of the breed referred to as a California Variegated Mutant (CVM). Regardless of colour, the entire breed holds the same characteristics.

▲ Romeldale ram meandering over to say hello at New Wave Fibre, Pender Island.

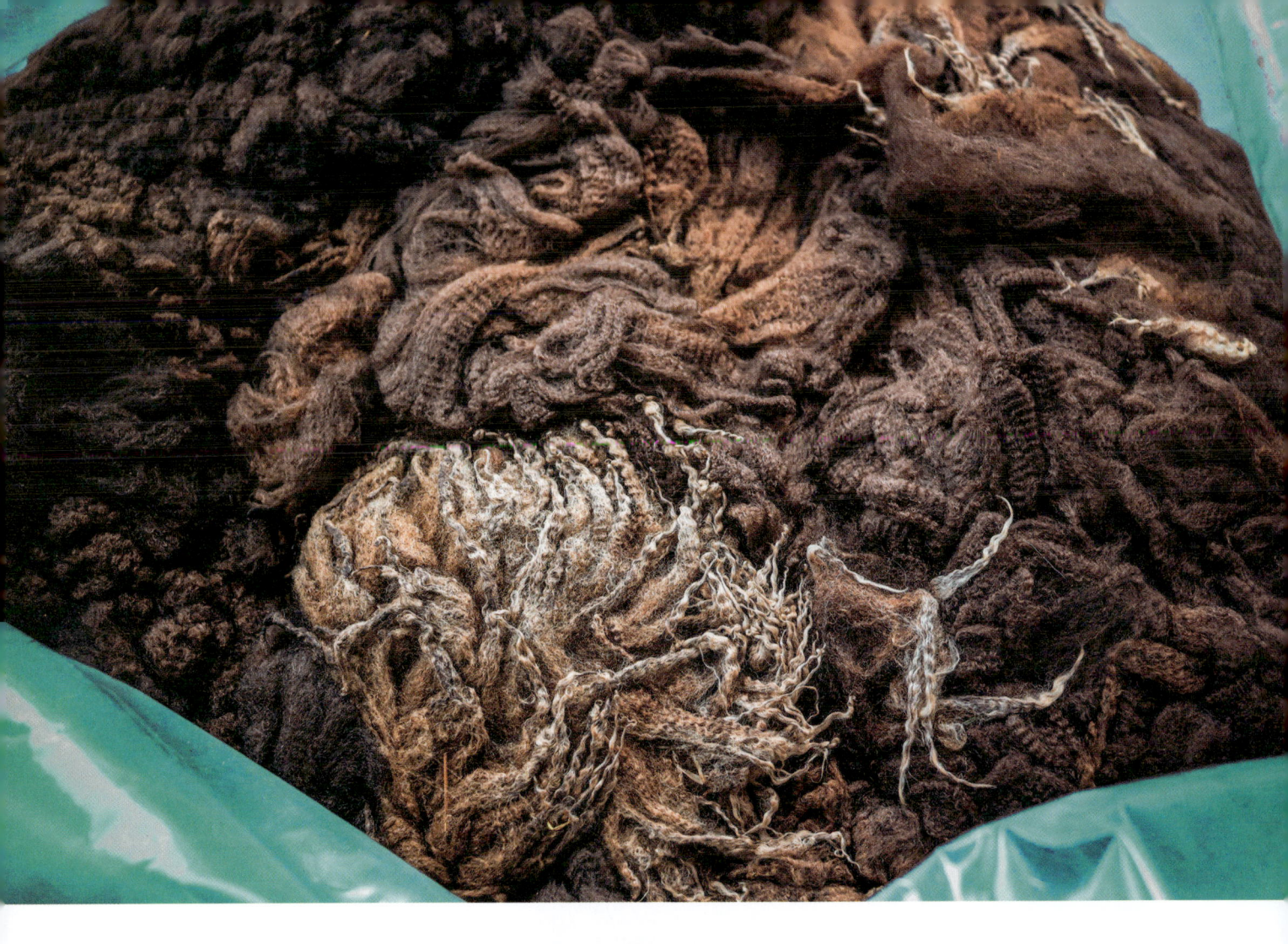

▲ Romeldale lamb fleece from New Wave Fibre, Pender Island.

► Romeldale-CVM ewe at New Wave Fibre, Pender Island.

The patterned fleeces are desirable for their range of colours. The fleece on coloured lambs often lightens with age.

USES: This fibre is desirable for next-to-skin and other fine garments, and is a good introductory variety for spinners looking to venture into finer wools. For knitters, this fibre will create a lofty, soft yarn. It is also suitable for felting. The white fleece will take on colour with a matte finish and the coloured fleeces can be over-dyed for some interesting results.

New Wave Fibre, Pender Island

(See entry under Black Welsh Mountain, page 33.)

Romney

This large breed is well known for its wool production and its calm, friendly disposition. The lustrous longwool is extremely versatile, with a staple length of four to eight inches. Longwools are a class of sheep encompassing many breeds that have a longer staple length and often a coarser fibre than a Down type. This breed can be shorn twice a year, which produces shorter staple lengths. The wool ranges from coarse to fine. The fleece is large and dense with a well-defined crimp (waviness). The raw fleece can have some kemp (coarse, straight fibre), a deciding factor when choosing your fleece. Flocks on the island contain white, grey, silver, black, and brown.

USES: Romney can be worn next to the skin but tends to be better as a second layer in a wide variety of wearable items like hats, mitts, jackets, scarves, and socks. It also works well for making durable home textiles like rugs, mats, and upholstery fabric. The fibre can be fun to process because it can be done in several ways with different results. Romney is a great fibre for anyone wanting to process fleece for the first time. You can card, flick, comb, or spin from clean fleece. Because this breed has low grease content, it can sometimes be spun fresh from the shear without having to wash it.

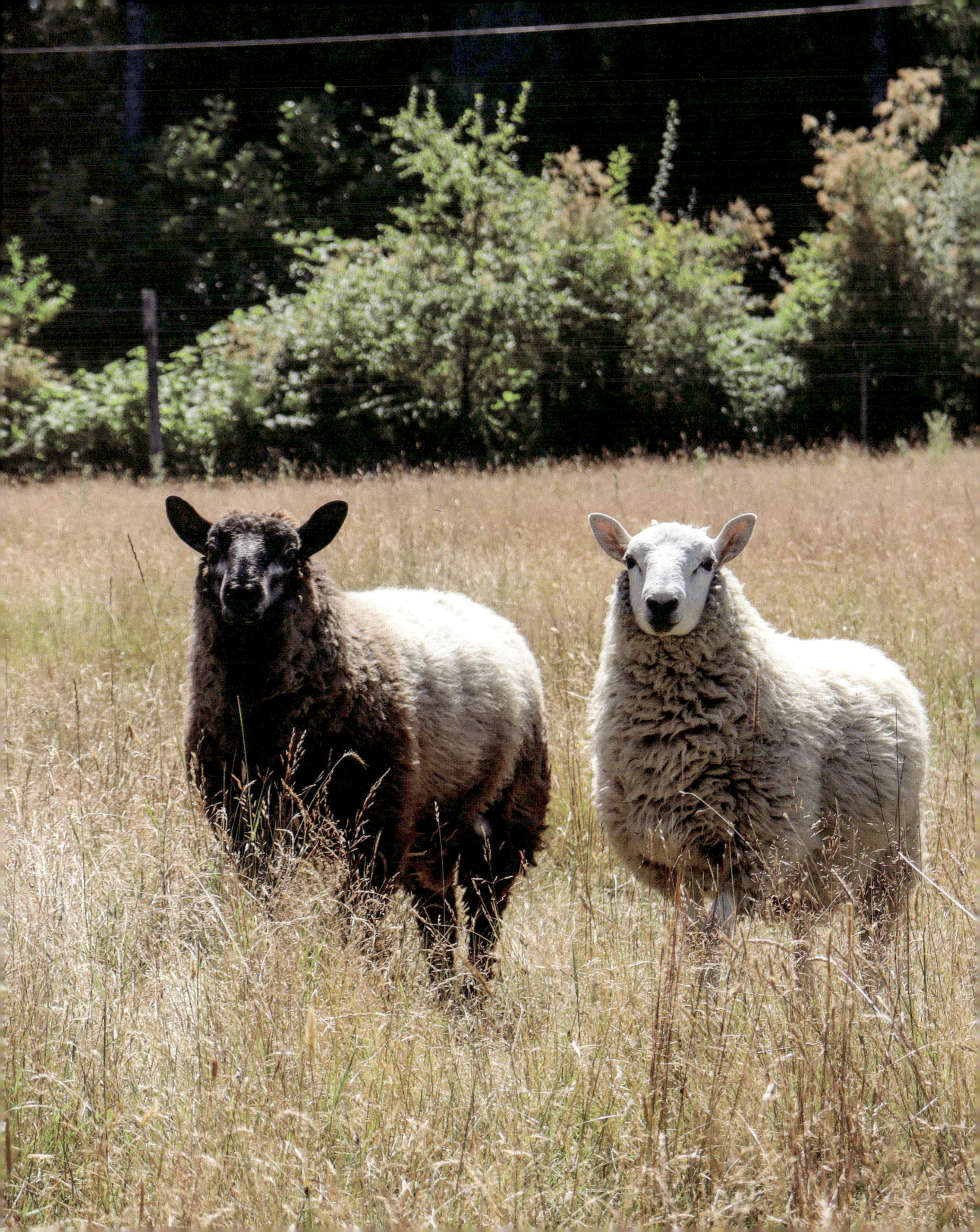

◀ Romney X ewes at Checker Grass Farm.

The white fleece will take dye colour well and the other colours can be experimented with for over-dyeing.

Checker Grass Farm, Errington
(See entry under Bluefaced Leicester, page 42.)

Fir Hill Farm, Pender Island
(See entry under Border Cheviot, page 54.)

Gun Barrel Canyon Alpaca Boys, Nanaimo
(See entry under Black Welsh Mountain, page 33.)

Salt Spring Island Wool Co., Salt Spring Island
Susan Astill runs Salt Spring Island Wool Co. Her mother used to operate the business, but Susan has taken over since her mother's passing. The business has operated for over fifteen years. The fibre comes mainly from Susan's brother's flock of sheep at Cottonwood Farm on Salt Spring Island. Cottonwood has around twenty Romney × Cheviots in natural colours like white, black to grey, and brown. Susan explains that the Romney can get a bit coarse, so they bred in some Cheviot to add softness while still retaining staple length.

The sheep are usually sheared at the end of May. Susan sends a large number of fleeces to Custom Woolen Mills to be processed. She then hand dyes some of the wool and creates one-of-a-kind painted skeins.

PRODUCTS: Two- and three-ply yarn, softspun yarn, hand-dyed and -painted skeins.

Swallows Keep, Galiano
(See entry under East Friesian, page 65.)

Windsor Farms, Salt Spring Island
(See entry under Bluefaced Leicester, page 42.)

Top This bottle-fed mixed breed was happy to come say hello at Checker Grass Farm.

Bottom Romney X at Windsor Farm.

Shetland

The first Shetlands were imported to Canada in the late 1970s. This breed is compact, hardy, and well suited for Vancouver Island and the Gulf Islands. They are friendly and easy to handle, which makes them a great family farm breed. Shetlands are one of the most colourful breeds, found in eleven colours and over thirty different colourway and pattern combinations, each with its own name, like "gulmoget," which means to have light underparts with a dark body, and "katmoget," which is the opposite, with light-coloured body and dark belly, legs, and face markings. Their fibre is silky soft yet durable, with a staple length of three to six inches.

Shetlands can be single- or double-coated. If they have a second outer layer, these guard hairs are usually not as soft as the undercoat. Shetlands also tend to grow an uneven fleece, meaning the fibre is finer and crimpier toward the neck and gets less fine, with smaller crimps, toward the back of the animal. This can make a single fleece very versatile, giving a maker many options for preparation and use.

USES: Shetland fibre is suitable for making a large range of items like scarves, hats, mitts, shawls, rugs, pillows, tablecloths, and table runners. The white wool will take on dye colour nicely and the large range of natural colours makes the options endless. The fibre is soft in the hand and the bouncy crimp spins easily, making it great to work with for spinners.

Left Well-trained border collie guiding the Shetland sheep on command at Guthrie Farm.

Right Shetland ewe at Root Spell Shetlands.

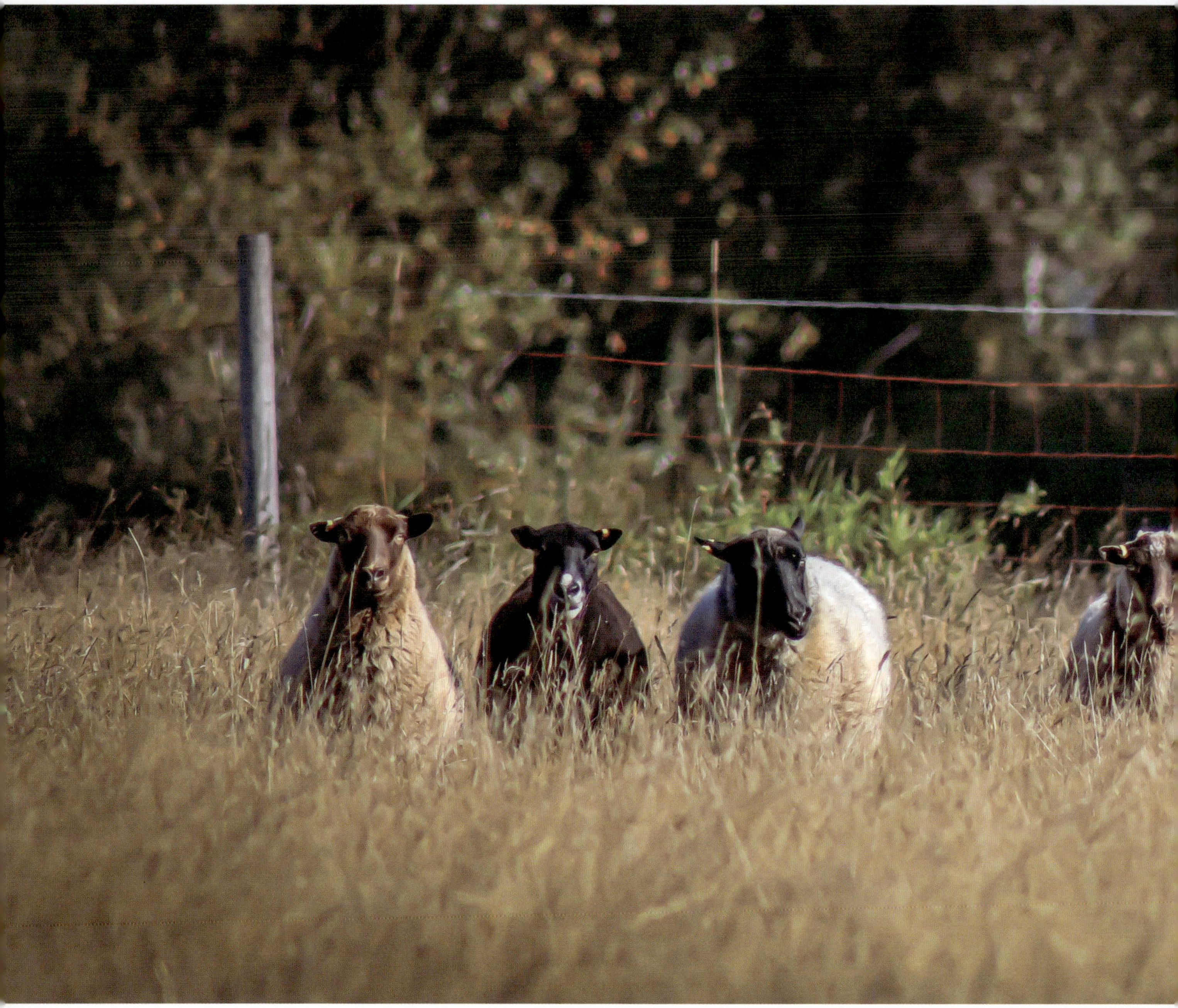

▲ Alert Shetland ewes in the pasture at Guthrie Farm.

Top White Shetland ram at Root Spell Shetlands.

Bottom Shetland lambs at Root Spell Shetlands.

▶ Shetland ewe with her two lambs.

Guthrie Farm, Merville

Anne Guthrie lets her small female border collie into the field and signals her to "look." The sheep all stand in a soldiered row, ears up, looking for her too. She spots them and, moving swiftly and low to the ground, she brings them in on Anne's command. The sheep listen with ease. The dog gives them space and holds off on the signal.

Anne has lived on the fifteen-acre property since the early 1990s, and in the Comox Valley for over fifty-five years. The white home, with its small back porch overlooking the orchard, is over one hundred years old. Anne tells me there used to be a massive three-storey barn as well, built on blocks. When the ground started to soften, it slowly started to tilt. Instead of destroying it, they sold it, and the buyer disassembled it piece by piece, numbering and labelling each chunk of wood. They took it away to be relocated, and Anne and her husband went to see it once it was rebuilt on the new property.

As the sheep approach the fence, I can see how soft and fluffy their reddish brown, white, and light grey fleece looks. There is one Black Welsh Mountain sheep in the mix. Anne explains that her Shetlands may be a bit on the larger side because she isn't breeding them, so they put that extra energy into their own body growth.

We go up to the house and Anne shows me some of the fleece she has washed and readied for spinning. She spins, weaves, and knits, often just spinning the washed fibre without combing it. The white fleece she shows me is truly white and would take dye colour well. Another is a light caramel.

Anne aims to have the flock sheared every eight to ten months.

PRODUCTS: Washed fleece.

Root Spell Shetlands, Nanaimo

Crouched in the tall pasture grasses, Ea Fable tells me about the Shetlands she's been raising since 2019. The ewes and lambs graze on the lower pasture of the sloped property, while the three brown, black, and white rams are in their own paddock off to the side.

The lambs get curious and come over to check me out as we sit. They are super soft, with dark black fleece. Ea tells me she is breeding for fibre, so she really pays attention to the fleece. She sends samples away to be micron tested and only breeds the sheep with the best fibre. She also watches that her crimp is consistent throughout the fleece. "I don't want the crimp falling off the back." This refers to a fleece that has a nice crimp near the chest and shoulders but loses the defined crimp toward the back.

Two of her ewes have the gorgeous, mottled fleece of the gulmoget colouring (light underparts with a dark body), which she is hoping to breed into the next generation of lambs.

The flock is shorn once a year, providing a fleece with approximately a four-inch staple length. The fibre is fine, with no guard hairs, suitable for spinning fine yarn and making soft garments.

PRODUCTS: Raw fleece.

Southdown

The Southdown family includes three sizes: toy or miniature Southdown being the smallest, Babydoll (see page 27) being the middle size, and the largest simply called Southdown. The breed is friendly and easy to handle. The larger Southdown is often raised for meat, whereas the other types are often raised for their fibre or just as farm pets. While there are some coloured Southdowns, they are more often white. The staple length is two to three inches, with dense, rectangular locks and a crimp that is often small and hard to distinguish.

USES: What you make with Southdown fibre will depend on the quality of the fleece. If you find yourself with coarser wool, stick to making durable items that aren't worn next to the body, like rugs and other home textiles. If the wool is soft and less coarse, it can be used for hats, mitts, socks, and sweaters. The wool is neither lustrous nor flat and will take dye colour nicely.

Gun Barrel Canyon Alpaca Boys, Nanaimo
(See entry under Black Welsh Mountain, page 33.)

Sunset Farm, Salt Spring Island
This twenty-three-acre farm is home to goats, chickens, ducks, and a mixed flock of over one hundred sheep, with three purebred rams that are Southdown, Suffolk, and Dorper.

Sandy Robley has owned and operated the farm for forty years and has worked with sheep for longer. She explains that Salt Spring Island is a great place to raise sheep because it doesn't have the predators found on the Mainland or even Vancouver Island.

Her flock is sheared in November. While this might seem odd, the sheep lamb in January and this timing keeps them closer to the barn, near their lambs.

The fleeces are made into a variety of products that are sold in the farm's studio, which is open every day of the year except Christmas.

PRODUCTS: Single-ply, two-ply, and eight-ply wool yarn, sheepskin slippers, wool socks, wool blankets, wool pillows, comforters, and sheepskins.

Suffolk

First brought to Canada in 1888, this breed was created by crossbreeding Norfolk Horn Ewes and Southdown rams. Because the breed has a reliable conversion of feed into meat, it is now the most common breed in North America.

The fleece, however, is often overlooked, though some breeders are producing desirable fleeces from their Suffolk flocks. The dense,

Top, left Young Suffolk ewe at Tsolum River Ranch.

Top, right Purebred Suffolk ram at Tsolum River Ranch.

Bottom, left Suffolk X being sheared at Parry Bay Farm.

Bottom, right Flock of Suffolk sheep grazing at Tsolum River Ranch.

▶ Texel ewe relaxing.

blocky locks are shorter, at two to three-and-a-half inches, and can be hard to distinguish from one another. Most Suffolk are found in white.

USES: This fibre will take dye colour, but with little lustre. It will be easier to card the shorter fibres and comb the longer ones. If you choose to comb the shorter fibres, you may experience some loss. The yarn spun from this wool is bouncy and lightweight, yet warm and durable. Suitable for making socks, hats, mitts, and sweaters.

Country Wools, Sidney
(See entry under Charollais, page 55.)

Parry Bay Farm, Metchosin
(See entry under East Friesian, page 65.)

Ruckle Farm, Salt Spring Island
(See entry under North Country Cheviot, page 89.)

Tsolum River Ranch, Courtenay
(See entry under Gotland, page 71.)

Windsor Farms, Salt Spring Island
(See entry under Bluefaced Leicester, page 42.)

Woodley Range, Ladysmith

This farm is nestled near the Woodley Range Ecological Reserve in Ladysmith. Beryl and Jody Shupe have been raising sheep since 2006, when they bought their first three sheep for their Australian cattle dog. The flock quickly grew and is now close to thirty-five sheep, made up of Suffolk and Suffolk × Texel.

Jody explains that they have sent their fleece away to be processed in the past, but the cost and the time it takes haven't always proven worthwhile. The farm has had some success selling raw fleece to artisans, and teachers have purchased fleece for teaching processing in schools. They have also had people buy raw fleece for insulating campers and making futon cushions.

Jody's father has gotten creative and used the seconds for lining nesting boxes and mulching the gardens.

The sheep are sheared at the end of May or beginning of June. During sorting, the young lambs' fleeces are kept separate from the older ewes because they are often softer and can make finer products.

PRODUCT: Raw white fleece.

Texel

The Texel was first brought to Canada in the early 1980s. A cross of Lincoln and Leicester Longwools, this breed is considered a high-quality meat-producer, so the fleece is often neglected.

▲ Valais Black Nose lamb (F2) at Gun Barrel Canyon Alpaca Boys.

▶ Young Valais Black Nose at Gun Barrel Canyon Alpaca Boys.

However, the matte, dense fleece is of a good quality, with a staple length between three and six inches, ranging from fine to coarse. It can be used to make many items. Watch for kemp (coarse, straight fibre) when choosing your fleece.

USES: When working with the fibre, try to find a fleece that is in the mid-range, neither fine nor coarse. This way, you have options when it comes to preparing it. If you are buying from a breeder who is growing primarily for meat, you might need to give the fleece extra attention when cleaning it. Combing can work well for removing extra vegetable matter. Because of the range from fine to coarse, this wool can be used in a variety of applications. It will take dye colour with low lustre.

Country Wools, Sidney
(See entry under Charollais, page 55.)

Parry Bay Farm, Metchosin
(See entry under East Friesian, page 65.)

Ruckle Heritage Farm, Salt Spring Island
(See entry under North Country Cheviot, page 89.)

Woodley Range, Ladysmith
(See entry under Suffolk, page 111.)

Valais Black Nose

This large-framed breed originated in the Valais region of Switzerland. Both males and females are horned, often white and curled, possibly with some black in them. They have black noses, ears, ankles, knees, and hooves. Females often have a spot on their tails. Their curly fleece hangs down over their eyes and they are often considered "super cute." They have very friendly personalities and are social with humans.

Though the breed cannot be imported to Canada, the frozen semen can be brought in to artificially inseminate a foundation ewe, preferably from a breed similar in size and characteristics to the Valais, such as the Scottish Blackface. In the first breeding season, this cross would result in an F1 lamb, which would be 50/50. The following year, you would breed one of the 50/50s with a different bloodline of Valais and have an F2, which would be 75 percent Valais, and so on. It would take five seasons of breeding before you would have a 97 percent Valais lamb.

Farmers looking to breed this sheep are looking for the challenge and are interested in working with a new breed. Valais are usually bred for their protein as well as their fibre; the fibre is known for making durable, long-lasting rugs, and is also good for felting. The breed should be shorn twice a year, averaging a six-inch staple length.

Gun Barrel Canyon Alpaca Boys, Nanaimo
(See entry under Black Welsh Mountain, page 33.)

▲ A Scottish Blackface ewe with her Valais Black Nose (F2) lambs.

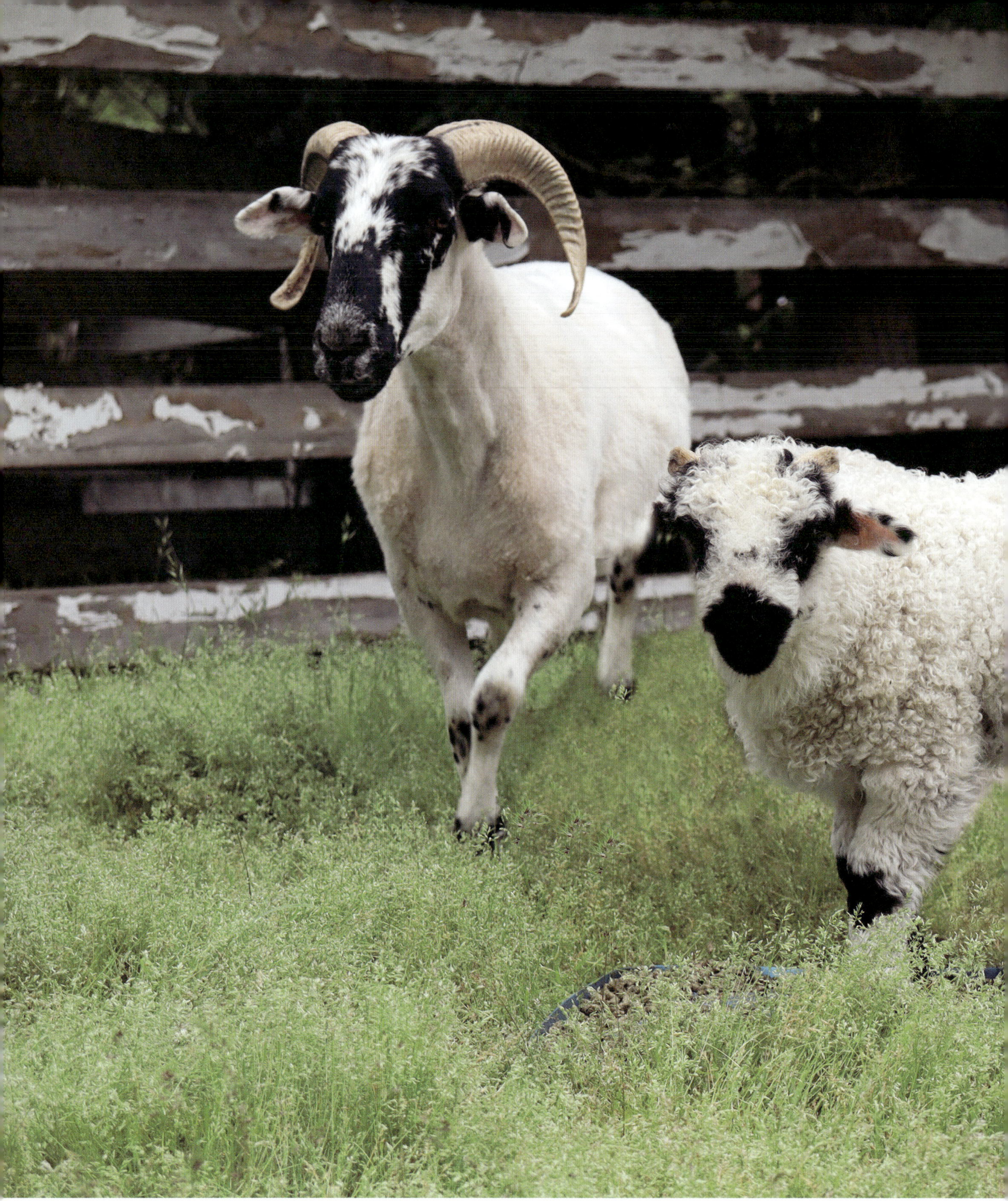

◀ A Scottish Blackface ewe with her F2 Valais Black Nose lamb at Gun Barrel Canyon Alpaca Boys.

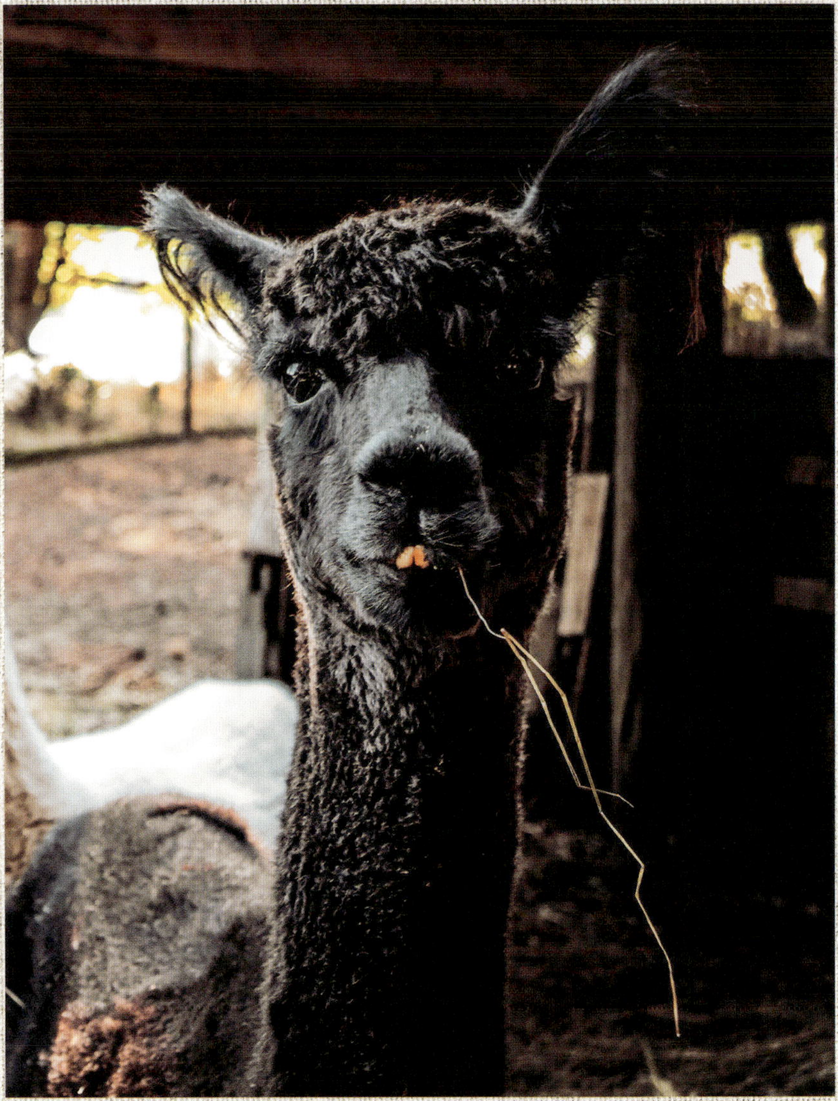

2 Alpaca

◀ Male alpaca at Yellow Point Alpacas.

ALPACAS (*VICUGNA PACOS*) are in the camelid family, originating in the high Andes in South American countries that include Chile, Peru, Ecuador, and Bolivia. They were first brought to North America in the late 1980s. According to the Canadian Llama and Alpaca Association, in 2006 Canada had 16,373 alpacas registered, and that number has been steadily increasing.

Alpacas are intelligent, friendly, clean, extremely hardy, and well suited for Vancouver Island and the Gulf Islands. As herd animals, they are happiest and healthiest living in groups. The average alpaca stands at thirty-four to thirty-eight inches in height and can weigh between ninety-five and two hundred pounds, depending on gender and variety. Alpacas do not need a whole lot of land to live on; a one-acre productive pasture can feed up to eight alpacas. If you get close to a herd, you can hear them communicating with one another through a series of throaty hums.

Interestingly, Alpacas tend to have poo piles, designated spots where they go to the washroom, rather than spreading poop around. If the flock has both males and females, they will tend to use separate piles. This natural instinct makes alpacas one of the tidier creatures.

Alpacas are raised mainly for fibre production. Similar to sheep, they are shorn annually. Networks like Alpaca Canada have established guidelines for shearing alpacas; these standards are important for helping the alpaca fibre industry develop and prosper.

▲ Alpacas at Hinterland Yarn.

Alpacas come in a wide range of colours from white to shades of grey and brown, almost red, and black.

There are two breed types that produce quite different fibre.

HUACAYA: The most popular breed is known as the woolly type, even though the fibre is not comparable to sheep wool. The Huacaya's face is fuzzy and teddy bear-like. The body shape is identical to the Suri, but the fibre grows off the body perpendicular and has a crimp, making the animal look fluffy. The staple length is two to six inches. The fibre is strong, resilient, and naturally water repellent. The durability juxtaposed against the soft texture of this fibre makes it extremely desirable.

SURI: Long, ringlet-type locks hang off the Suri's body in a variety of textures, with a staple length up to eleven inches. The Suri is sometimes only sheared every two years. The fibre is very smooth, high in lustre, and silky to the touch. Because this fibre has no crimp, there is no elasticity and the yarn will not bounce back. Makers need to consider their gauge or sett when working with alpaca to prevent the finished product from sagging. If you make your garment with too tight a gauge/sett, it is likely to be stiff; too loose and the yarn can shift around and deform.

USES: For added elasticity, you can blend alpaca with high-crimp wool like Shetland or Rambouillet. Other interesting blends are with silk, Angora, and mohair. When spinning this fibre, keep in mind that the long, slippery locks might need more twists to hold together, but they can easily become over twisted, creating wiry yarn. The use of carders with fine teeth is recommended to prepare the fleece for spinning. The fibre's high lustre will make dye colour pop.

Brook Meadows Farm, Saanichton

Martha and Peter Klinovsky run this alpaca farm. They shear their animals in June and the fibre is mostly black.

PRODUCTS: Raw alpaca fibre.

Hinterland Yarn, Pender Island

(See the main entry on Hinterland Yarn under Bluefaced Leicester, page 42.)

Hanahlie Beise takes me to visit her flock of neutered male alpacas, explaining that she got a few from a retirement flock and the rest are rescue animals, many surrendered to the SPCA. Some of them are getting on in age, but these adorable, goofy creatures are all happily munching on fresh hay under a beautifully designed and built lean-to structure.

The flock was sheared in the spring, so their fleece shows only a few months of growth when I visit. Each one has a unique face and fleece. There is one llama in the flock that wasn't sheared this season, so his coat is long and fluffy.

Hanahlie shows me all her Watershed, Range, and Dusk yarn lines, blending undyed Canadian Rambouillet and her alpaca fibre. The colours are natural and complementary to our landscape, and the fibre is soft yet durable. She uses Maiwa natural dyes, and the colours she produces are vibrant and make me smile.

PRODUCTS: Woolen-spun single-ply chunky, woolen-spun two-ply worsted, single-ply woolen spun fingering, knit kits, patterns, duvets and pillows, and sheepskins.

Pacific Sun Alpacas, Duncan

If you need to know anything about alpacas, Jenny Young is the woman to ask. She has been breeding alpacas since the late 1990s on her five-acre farm and has taken many courses on fibre handling and testing. When her four daughters were young, they had horses on the property, but as the girls grew up, the horses faded away and Jenny found her passion for alpacas.

We are looking at the fibre from the most recent shear. "Here," she says, "hold the fibre up to the light. See how when you pull a little away it looks as fine as cobwebs? That's a trick," she explains, "for knowing if your micron count is below twenty-eight."

▲ A male alpaca at Yellow Point Alpacas.

▶ Freshly shorn alpacas grazing at Pacific Sun Alpacas.

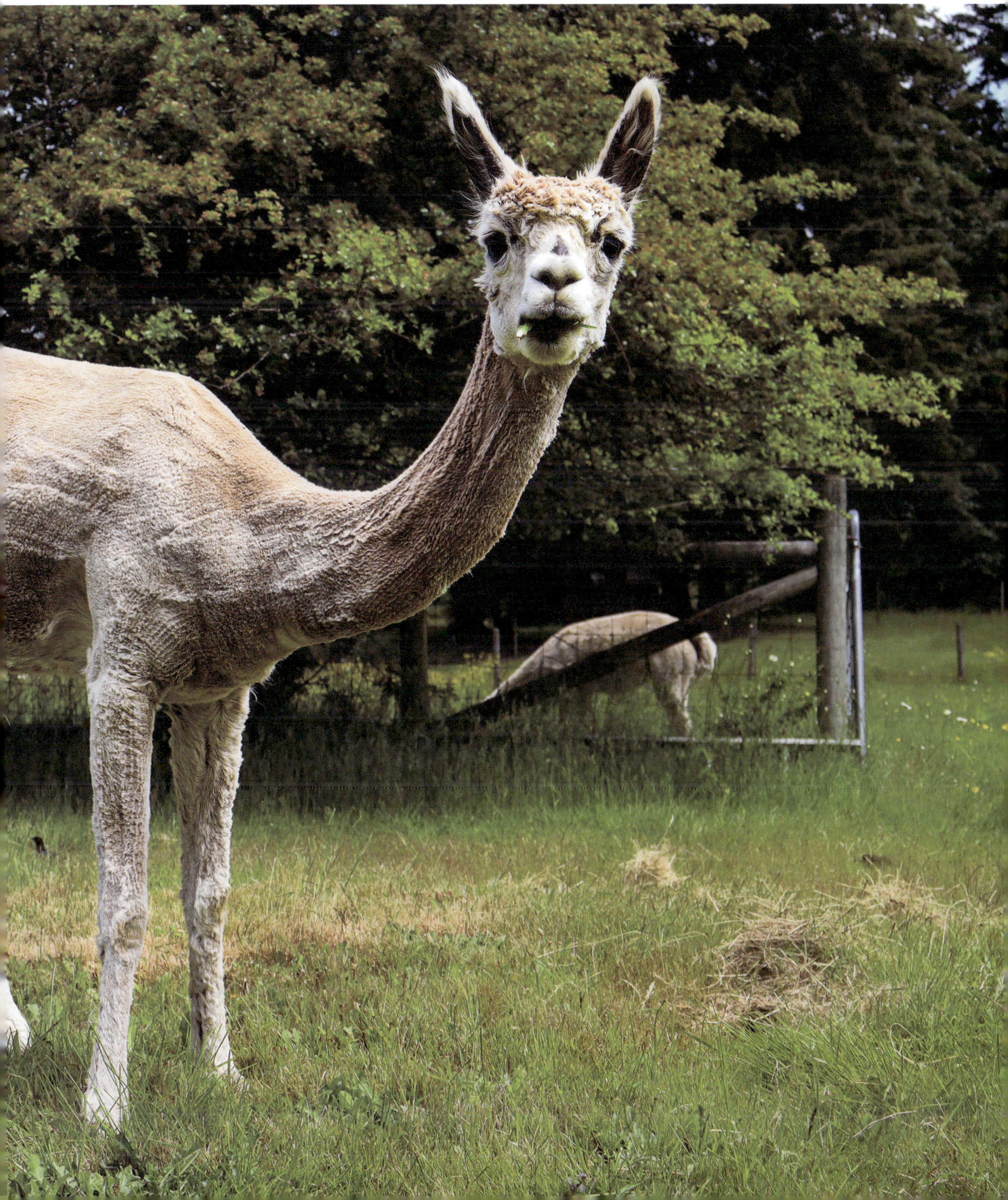

Jenny chose to breed alpacas because she believes they are the most eco-friendly livestock species, being clean (their padded feet don't leave the ground dug up and muddy), gentle on the infrastructure, and putting out manure with no strong smell that makes great fertilizer. Further, Jenny reminds me that alpacas are often eaten in other countries, making all their parts usable.

PRODUCTS: Roving in a variety of natural colours, raw fleece.

Sea Dog Farm, Saanichton

Shawn and Katy Connelly have transformed their five-acre property into a productive organic fruit, vegetable, cannabis, and flower farm. The property was run down when they acquired it, but friends and family have helped turn it into a dream. Together, they have brought the once-neglected orchard back to life, and planted more fruit and nut trees, plus raspberries and blueberries. With seventy raised beds and a forty-foot greenhouse, a lot of food and flowers are produced on this farm.

The property is also home to four female alpacas. Katy and Shawn tell me how the alpacas love to eat garden waste, like greens from bean and pea stalks, turning what would otherwise go into the compost pile into instant fertilizer. The alpacas have multiple areas to graze so they don't eat everything right down.

Shearing is in late May, before the heat of June hits. Though the farm supported the local mill when it was running, they now have to send the fibre off island to be processed.

PRODUCT: Alpaca roving, raw alpaca fibre.

Spring Valley Farms, Denman Island

Spring Valley Farms, run by Rose and Bruce Fell, is home to both llamas and alpacas. Rose washes and cards the fibre by hand; it is then locally spun and made into products. "It doesn't always pay much," says Rose. What keeps her processing is "the pleasure of working with the fibre and working from home."

Rose is a seasoned shearer, with over twenty-one years of experience shearing llamas, alpacas, and Angora and cashmere goats across Canada.

PRODUCTS: Raw llama and alpaca fibre, washed and carded fibre, 100 percent alpaca and llama mitts, socks, and toques. Llama and alpaca shearing services.

Sunhill Orchard, Victoria

This once commercial orchard is now home to thirteen alpacas and one llama. Cathy Christopher explains that she always wanted to own a farm. In 2010, that became a reality when she bought this property, which she has spent years transforming into her dream. The lush, productive gardens and well-manicured property show it is cared for. Some of the remaining Gravenstein apple trees in the orchard are still productive, and the farm grows and sells a large variety of fruit and vegetables at their farm stand.

The flock is friendly, and each alpaca has a unique fleece, with a large range of colours from nearly black to fawn, rose grey, silver grey, and brown. The flock is shorn once a year, and the fibre is sent to the east coast to be processed. Cathy also sends her seconds out to be processed into dryer balls. She explains that she supported the Inca Dinca Do mill when it was running, and would love to see another island mill start up.

The fibre comes back as ready-to-spin roving and blended two-ply.

The farm has a fibre studio that teaches weaving and knitting.

PRODUCTS: Alpaca roving, raw fleece, two-ply blended yarn, dryer balls, and a large variety of seasonal produce.

Yellow Point Alpacas, Ladysmith

Lenie and Peter Johnson have been running this oceanfront farm since 2007. After visiting alpaca farms on holidays, they fell in love with the animals, and they now have fifteen alpacas.

When I arrive, Peter greets me and we tour the farm store, which is full of cozy alpaca products. Next, we go out and visit the females, which are happily chomping hay and resting in the dappled light. Their colours are beautiful—brown, white, and black. The males are kept in a separate fenced area on the upper side of the store. Peter lures them closer with fresh spruce branches.

The flock was shorn at the beginning of summer and doesn't have a lot of growth in their fleece yet. As well, the mild fall has resulted in slower growth since, as Peter explains, fleece grows faster in cold weather.

You can visit the alpacas and the farm store on the Cedar & Yellow Point Artisans Tour in November.

PRODUCTS: Raw alpaca fibre, skeins, socks, mitts, hats, scarves, blankets, and sweaters.

▲ A young female alpaca, freshly shorn, at Pacific Sun Alpacas.

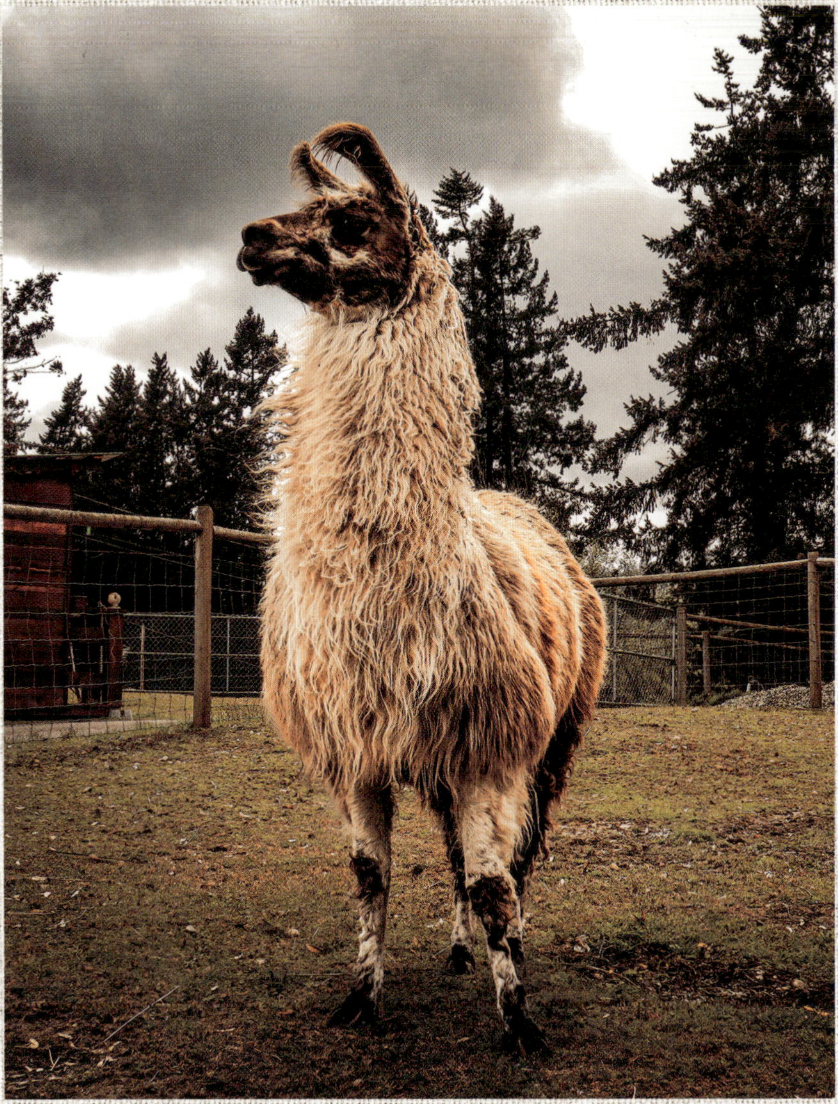

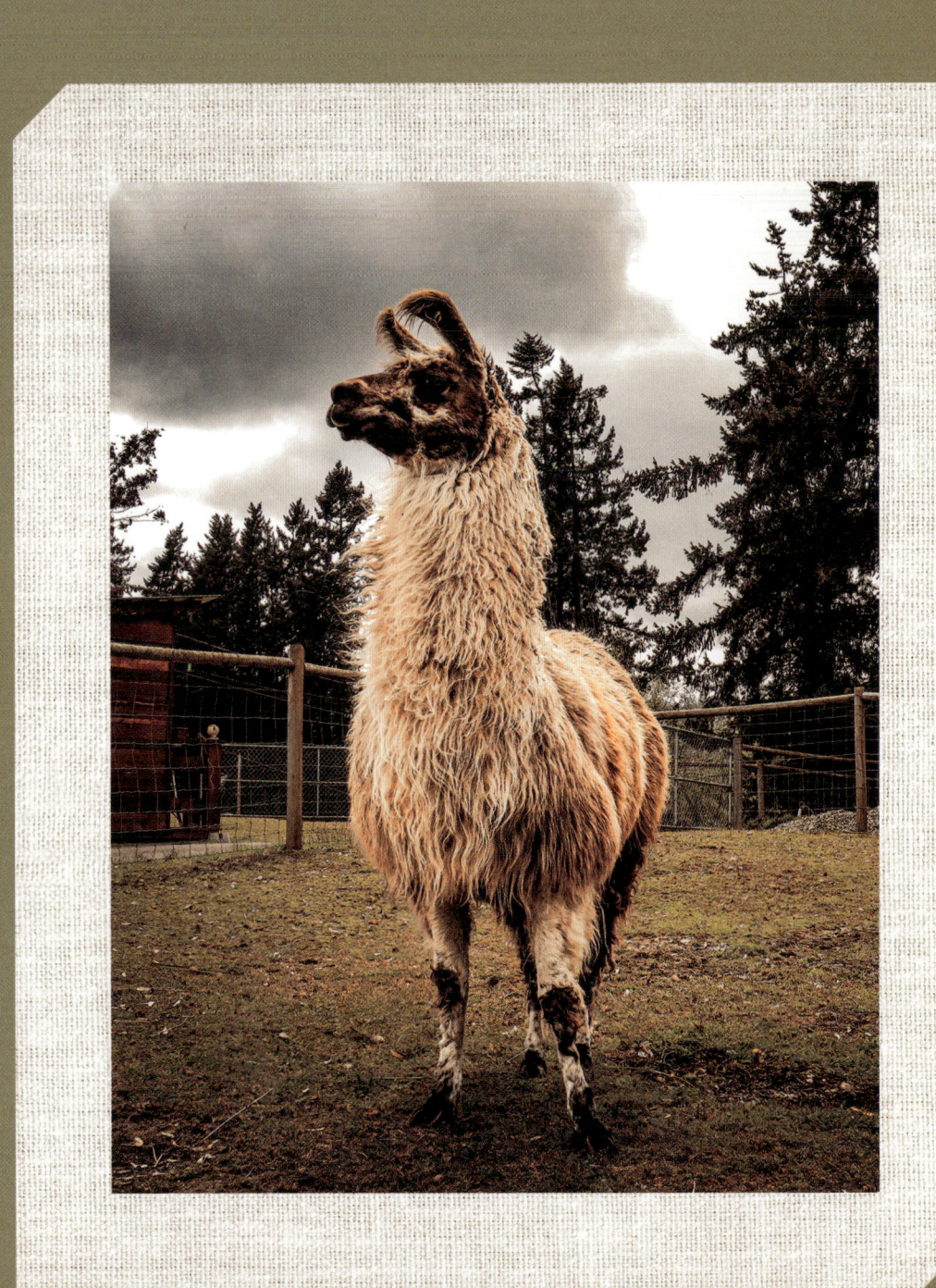

3 Llama

◀ A llama at Woosterville Mini-Llamas.

LIKE ALPACAS, llamas (*Lama glama*) are members of the camelid family and originate in South American countries like Bolivia, Peru, Ecuador, Chile, and Argentina. They were first brought to North America in the late 1800s and early 1900s. As herd animals, llamas like to be kept in groups or with other herd animals like sheep, goats, and sometimes horses.

Vancouver Island and the Gulf Islands are home to llamas in a variety of shapes and sizes, from mini-llamas to full-sized woolly and Suri types. I will focus on the woolly type because it is the most common type bred on the Island for fibre production.

Like the alpaca, the woolly type llama grows fluffy, slightly crimped fibre. They can come single or double-coated. The outer protective layer is coarse, crimp-less and referred to as the guard hairs. Guard hairs do not take on dye colour very well and create a more wiry, scratchy fibre. The insulating undercoat is much finer and usually a different colour than the guard hairs. The staple length of the undercoat is three to eight inches.

Llamas are unique because they shed, and their fibre can be gathered through combing; they do not always need to be sheared. However, most llama owners on the islands shear annually to keep the animals cooler through the summer months. Llamas love vegetable matter so their fleece can often be dirty, which can deter a processor from choosing llama fleece. However, llama is a fun, versatile fibre to work with.

▶ A llama at Millstream Miniature Llamas.

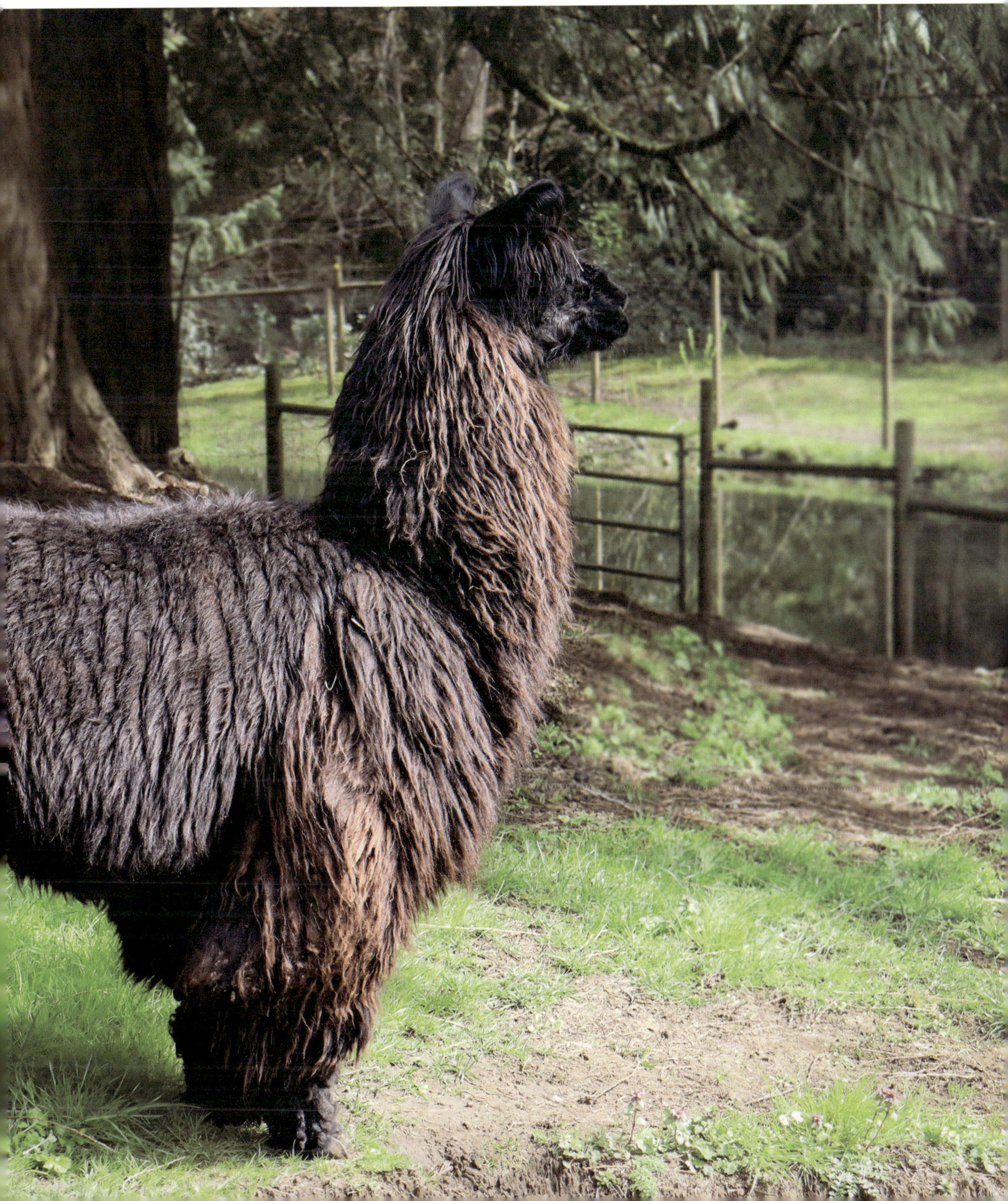

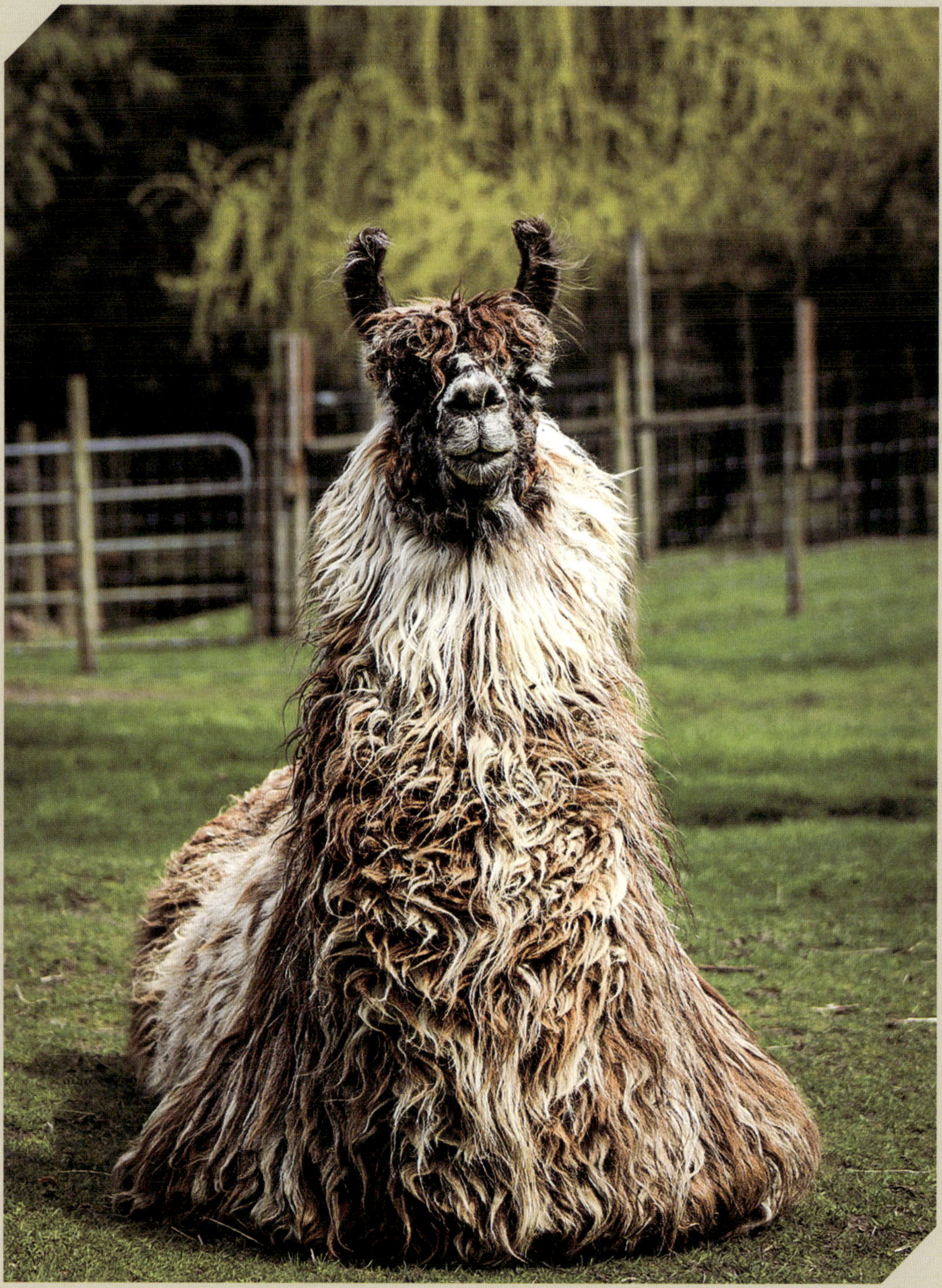

USES: Llama fibre is lanolin-free, so it does not usually need to be heavily scoured. However, thanks to the llamas' love for vegetable matter, the fleece may need to be thoroughly skirted and shaken to remove debris. Because llama fibre is warmer than sheep wool, you can spin a finer yarn and make a lighter garment while retaining the warmth of a chunkier wool garment. Llama yarn does not have as much elasticity as wool, so tension needs more consideration to avoid misshapen garments. Blending your llama fibre with wool can help create more elasticity and structural integrity in your final creation.

◀ Female llama at Millstream Miniature Llamas.

Millstream Miniature Llamas, Victoria

Turning onto the driveway of this farm and following the stream down to the house, I'm not sure how an acreage housing eighteen mini llamas can exist in this suburban area. But the five-acre farm is a secluded gem tucked well away from the main road. A large white farmhouse stands proud on the hill overlooking the property. The llamas graze in many different areas of the fields and have multiple outbuildings to choose from. The babbling stream that makes up the entire southern perimeter of the property gives the farm a rural feel.

Lavinia and Alan Stevens have owned and operated the farm since 1994. The farm had other uses, like a B&B, before they began focusing on mini llamas in 2010. They were one of the first Canadian farms to breed minis.

They processed their fibre locally when that was an option, but now they send it out of province. The farm store sells lots of fibre goods.

PRODUCTS: Llama roving and yarn, dryer balls, handmade llama figures, handmade hats, and mitts.

Marshall-Inman Farm, Sooke

Basil and Glenys Marshall-Inman have been breeding and caring for llamas since 1988. Their farm is home to four llamas, though in the past they have had larger herds. When the fibre is shorn, a local spinner and maker, Patricia Hilton, processes it into yarn and products, blending some fibre with merino to increase the quality.

Patricia hand knits the wool products that are sold in the farm's gallery store. The gallery also showcases Basil's woodwork and Glenys's pottery. Together, they have filled their shop with magical handmade creations. There are functional pieces mixed with sculptural ones, wooden toys, ceramic mugs, cutting boards, jewellery, and more.

PRODUCTS: Llama skeins, hats, mitts, scarves, socks, ceramic and wood products.

Woosterville Mini-Llama Farm, Duncan

When most people think about retirement, they don't think about upsizing and taking on a farm full of llamas. But Pat Morgan and her husband did just that. They bought two llamas before they even bought the farm in 2011. Pat fell in love with the llamas' big eyes and long lashes, and she's never looked back.

The farm currently houses sixteen llamas, some true miniatures and others standard-sized. There are some Suri types with long silky locks that drape more than fluff and an Argentinian male, Cracker Jack, that is fluffy from head to toe.

The farm washes and processes some of its own fibre. They use their carding and felting machines to make beautiful art, such as wall hangings and felted bowls. The fibre they don't process themselves gets shipped to the east coast and sent back as roving and yarn. Any fleece that isn't used as fibre is used in other creative ways, like lining hanging baskets and insulating the chicken coop.

PRODUCTS: Llama roving, yarn, hand-felted llama figures, felted decorations, and cards.

▲ Poco Rojo is a male Argentinian llama at Millstream Miniature Llamas.

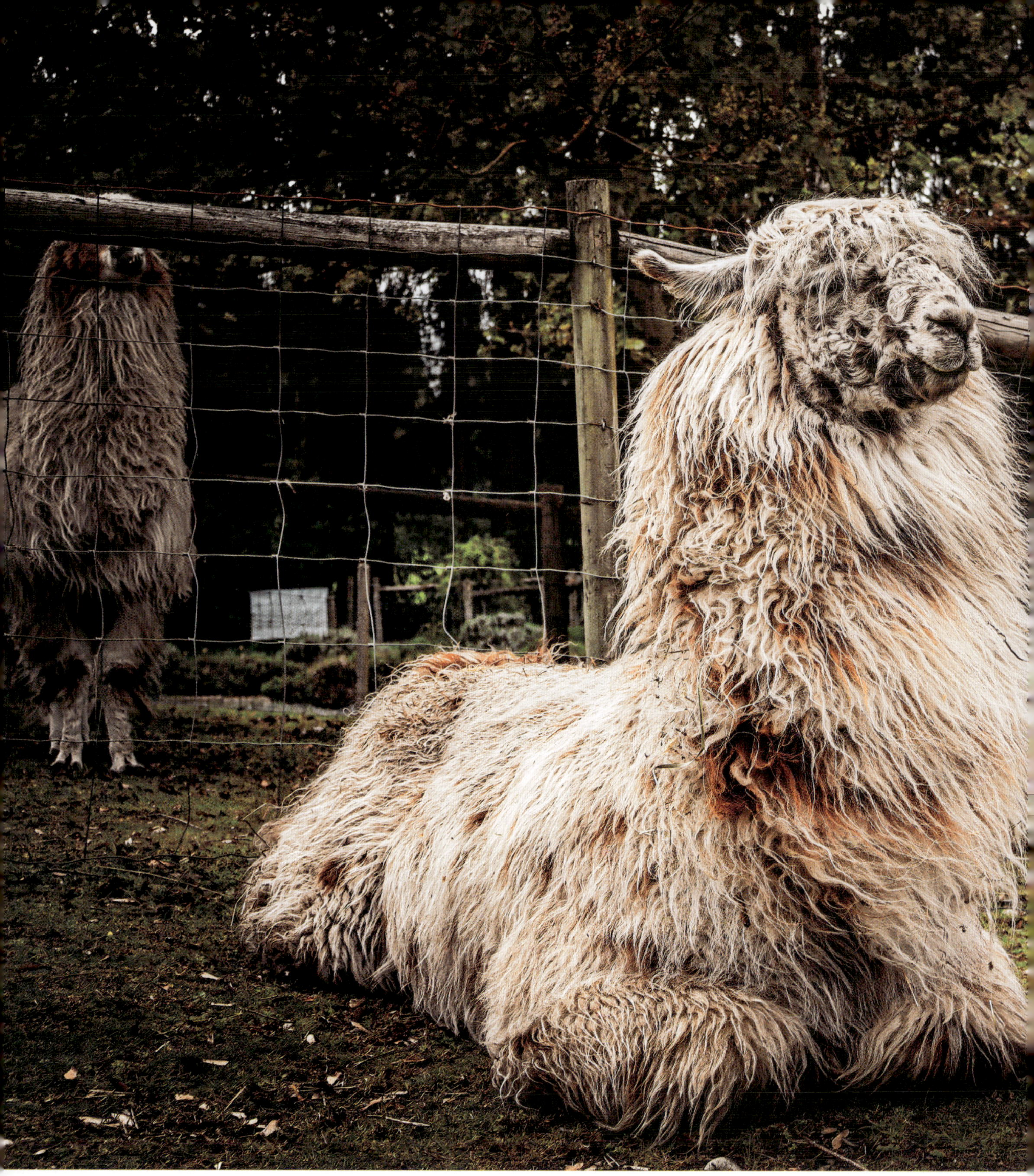

◀ Cracker Jack, the Argentinian male llama, at Woosterville Mini-Llama Farm.

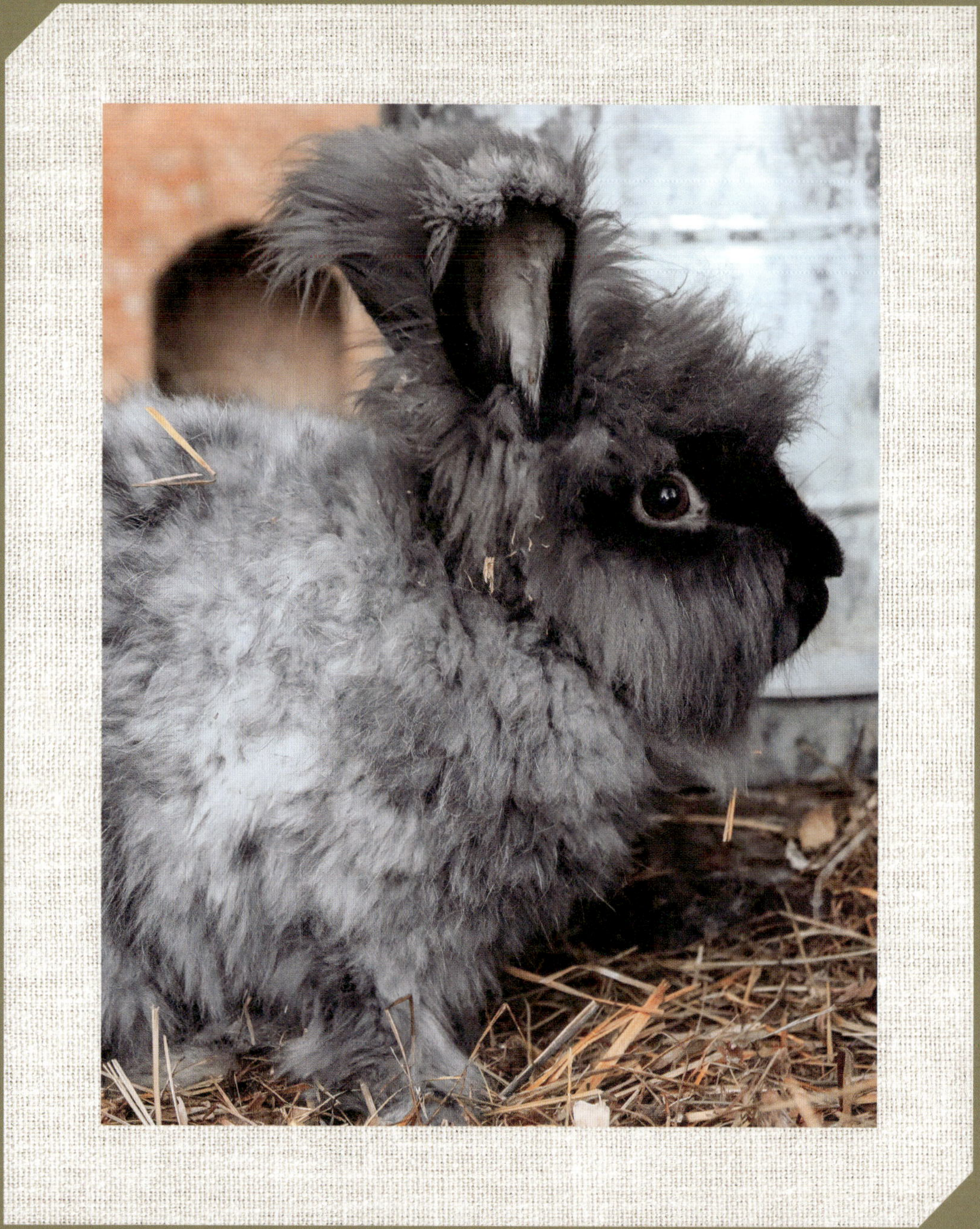

4

Angora Rabbit

◄ A grey Angora rabbit at Tsolum River Ranch.

ANGORA RABBITS (*Oryctolagus cuniculus*) are a special breed of rabbit whose hair grows extra long and fluffy. Their average weight is five and a half to twelve pounds. These rabbits are dependent on humans to comb and maintain their fibre so it doesn't get too long and unmanageable. They need a haircut or shearing every three to twelve months, when the fibre is three to six inches in length, depending on the harvesting schedule.

Angoras are unethically treated in the world textile economy. In large-scale Angora mills, fibre is removed through plucking. Animals are often hung by their feet and their fur torn from their bodies. This is unnecessary. Vancouver Island and the Gulf Islands have breeders who ethically breed and raise Angora fibre. They care for their rabbits like pets and use only humane methods for hair removal.

Angoras come in many varieties, including English, French, German, Satin, and Giant. Two active breeders on Vancouver Island and Salt Spring Island are working with German Angoras. Considered *the* fibre variety, German Angoras can produce up to five pounds of luxury fibre a year. They need less maintenance than most other breeds as their fibre is resistant to matting and they do not moult. The fibre will be better quality, however, if the animal is regularly combed. The undercoat does have a crimp, though it can't be compared to sheep wool. You can produce a yarn with a small bit of elasticity from the more crimped fibres, but Angora tends to be a drapey yarn. The fibre doesn't hold together like

▲ Grey Angora bunny at New Wave Fibre, Pender Island.

▶ Close-up of fibre from a white Angora rabbit at New Wave Fibre, Pender Island.

Opposite Kelsey Epp from Tsolum River Ranch giving her Angora rabbit a combing.

locks of wool; rather, it can tangle, with individual fibres going in all directions. When prepared properly, Angora fibre makes garments that are super soft, warm, and lightweight.

USES: Angora fibre is usually combed before it is cut or shorn from the animal and is easiest to handle when it is unwashed. If you want to spin pure Angora yarn, find the longest staple length you can. The fibre should not need carding, but if you want to separate the fibres slightly before spinning, use hand carders and very gently pull the fibres apart. If you are blending your Angora, a drum carder with a fine comb will work well. Angora blends beautifully with other fibres, adding its signature haloed look to the yarn. Remember that this fibre is more like hair and can be a bit messy. Use a designated work area for preparing and spinning so you do not get bits of fibre everywhere. This fibre is best spun fine with a good amount of twist to hold it firmly together, but it can become wiry if overspun. Try to spin the fibre cuticle end first, as it should catch and hold better, leaving the tapered tips to form a halo.

New Wave Fibre, Thetis and Pender Islands
(See entry under Bluefaced Leicester, page 42.)

Tsolum River Ranch, Comox Valley
(See entry under Gotland, page 71.)

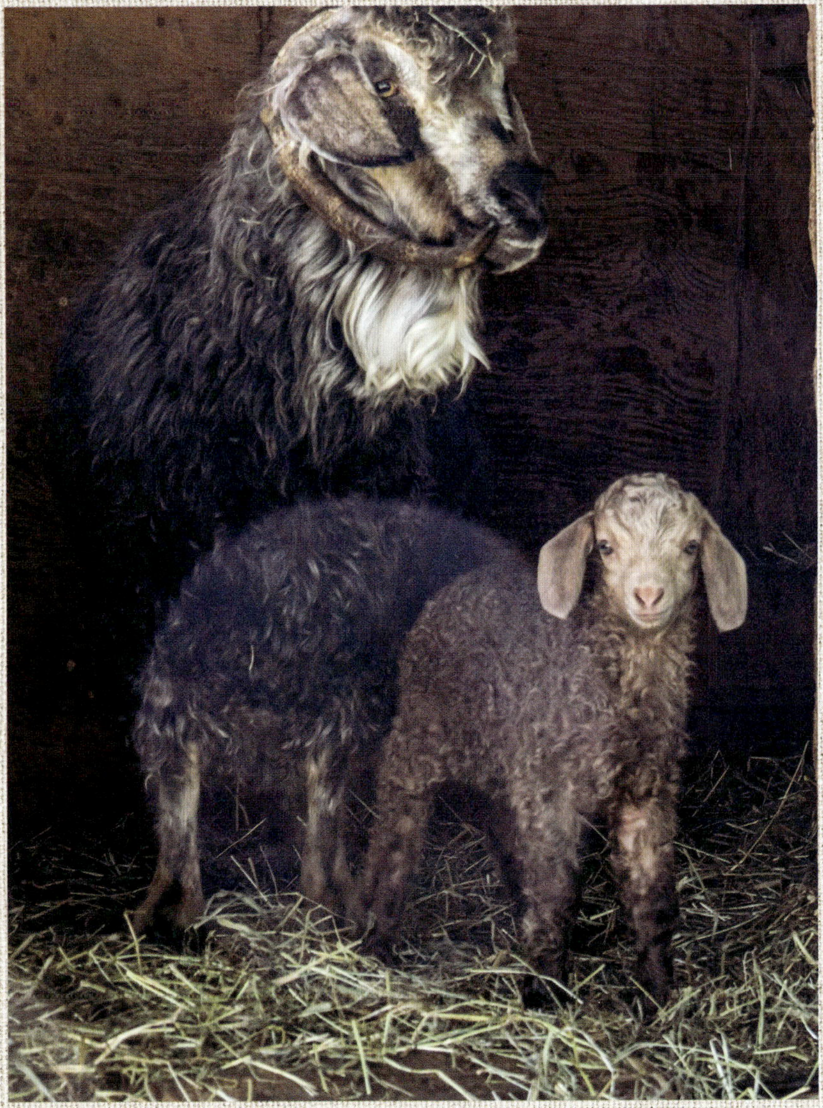

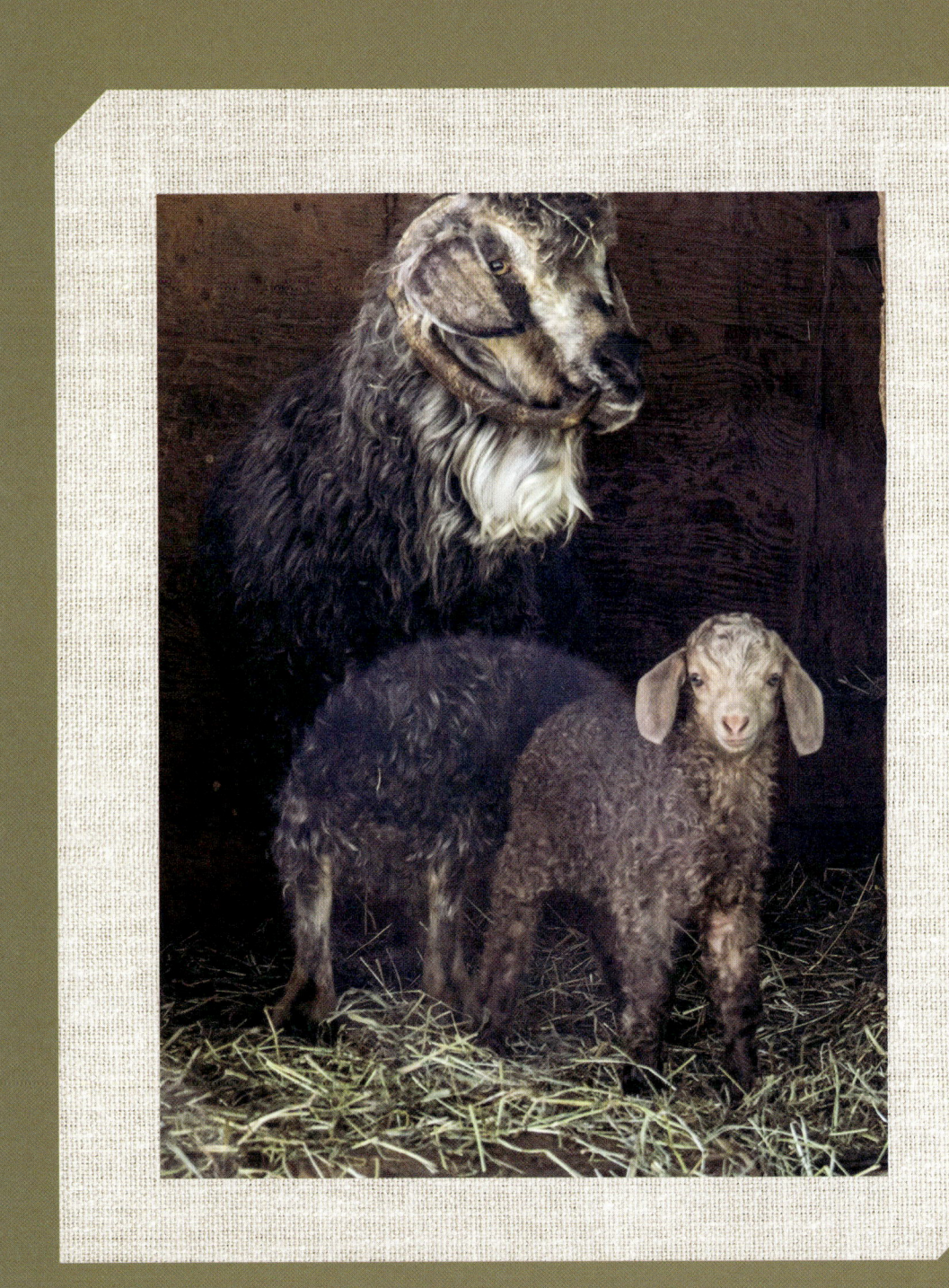

5 Goat

◂ Angora goats at Up A Creek Farm.

GOATS (*CAPRA HIRCUS*) are an important part of animal agriculture here on Vancouver Island and the Gulf Islands. These goats differ greatly from the mountain goats that were once used as a staple fibre here in our region; their genes are not mixed, though they do belong to the same biological family. The domesticated herd animals we use today have many uses. They produce high-protein meat, their milk is used to make cheese and soap, and their hides can be tanned and used to make a variety of useful items. Goats are great for cleaning up pastures or brambly areas. And, of course, they produce fibre.

Depending on the breed, goats have great personalities and tend to be social, though I've been told some can be grouchy, protective, or dominant, and use their horns more than needed. I will focus on the goats producing fibre in our region.

Cashmere

Cashmere refers not to a specific breed of goat, but to the downy undercoat that most goats produce (except for the single-coated Angoras). The undercoat is covered by another layer of guard hairs, which are thicker, straighter, and usually a different colour than the undercoat. To be classified as cashmere in Canada, the fibre must be nineteen microns or less and have a minimum length of 1.25 inches. There must also be a clear differentiation between the fibre and the guard hairs.

▲ A Spanish cashmere buck and doe at Reynolds Family Farm.

Opposite, left Cashmere fibre from Reynolds Family Farm.

Opposite, right Dark guard hairs found in the cashmere fibre.

Cashmere is considered a high-quality luxury fibre, desirable in the textile industry for many reasons. It can be blended with a lesser quality fibre to improve the texture and quality of the finished product.

The word cashmere is derived from *Kashmir,* a province in the Himalayas where the fibre is grown in the highest quantity and quality.

Goats producing cashmere-quality fibre began being bred in Canada in the late 1980s and early '90s. The genetics are said to come from Australia and the US. The most common breed types producing fibre in North America are pygmy, Spanish, and fainting goats. The cooler climates make for more undercoat production, as this is essentially the animal's winter coat.

Cashmere comes in a variety of colours: white, cream, brown, black, red, grey, and badger-faced. Cashmere-producing goats will begin to look fluffy and maybe messy when their undercoat needs to be groomed. The fibre grows heaviest on the goat's sides, back, neck, chest, and rump.

USES: Cashmere can be used to make luxury items, most often sweaters, cardigans, scarves, shawls, hats, mitts, jackets, and blankets. More delicate than woollen garments, cashmere garments will not stand up to the same hard use.

▶ The buck of the Reynolds Family Farm herd. The males' horns grow out to the side and make a large display.

The fibre is removed from the goat by combing at the right time of year. If the combing is done too early, the soft cashmere fibre won't come off the animal as easily. Combing too late, you are likely to hit the goat's big shed, when all the guard hairs are in the fibre. Early spring is a good time, but combing too early may leave the goats cold. Once the fibre is removed from the animal, the guard hairs that do get in need to be separated from the undercoat. If you are purchasing raw cashmere, you are likely to get some guard hairs. They are easy to tell apart from the cashmere fibre and can be easily removed with a little patient handwork.

Cashmere doesn't contain lanolin, and because it's the undercoat it is often fairly clean. This means it doesn't need to be washed before home processing. However, goats tend to smell like goat cheese and the fibre does hold some of that smell. Washing the finished garment is recommended, rather than trying to wash the raw fibre, which is likely to tangle and felt up. To spin your yarn, you can often work from a dehaired mass, making sure to handle your wool gently so it doesn't tangle and pill. If necessary, pull the fibres apart gently by hand.

The fibre is matte and lustre-free, but the whites, greys, and lighter shades will take dye colour and can produce some interesting results.

Reynolds Family Farm, Mill Bay

Cara Reynolds runs the only registered cashmere farm on Vancouver Island, in Mill Bay where the mild temperatures and ocean air are ideal for growing cashmere. The goats and chickens roam on this free-range farm, the open pastures surrounded by Douglas firs and giant maples. Each goat is called by name and has a personality of its own. Cara used to breed, but the demand was often for meat animals, and she aimed to breed these goats for fibre enthusiasts. At this time, she maintains her flock for fibre and companionship; mainly made up of a Spanish breed of cashmere goat, the flock has some other interesting characters.

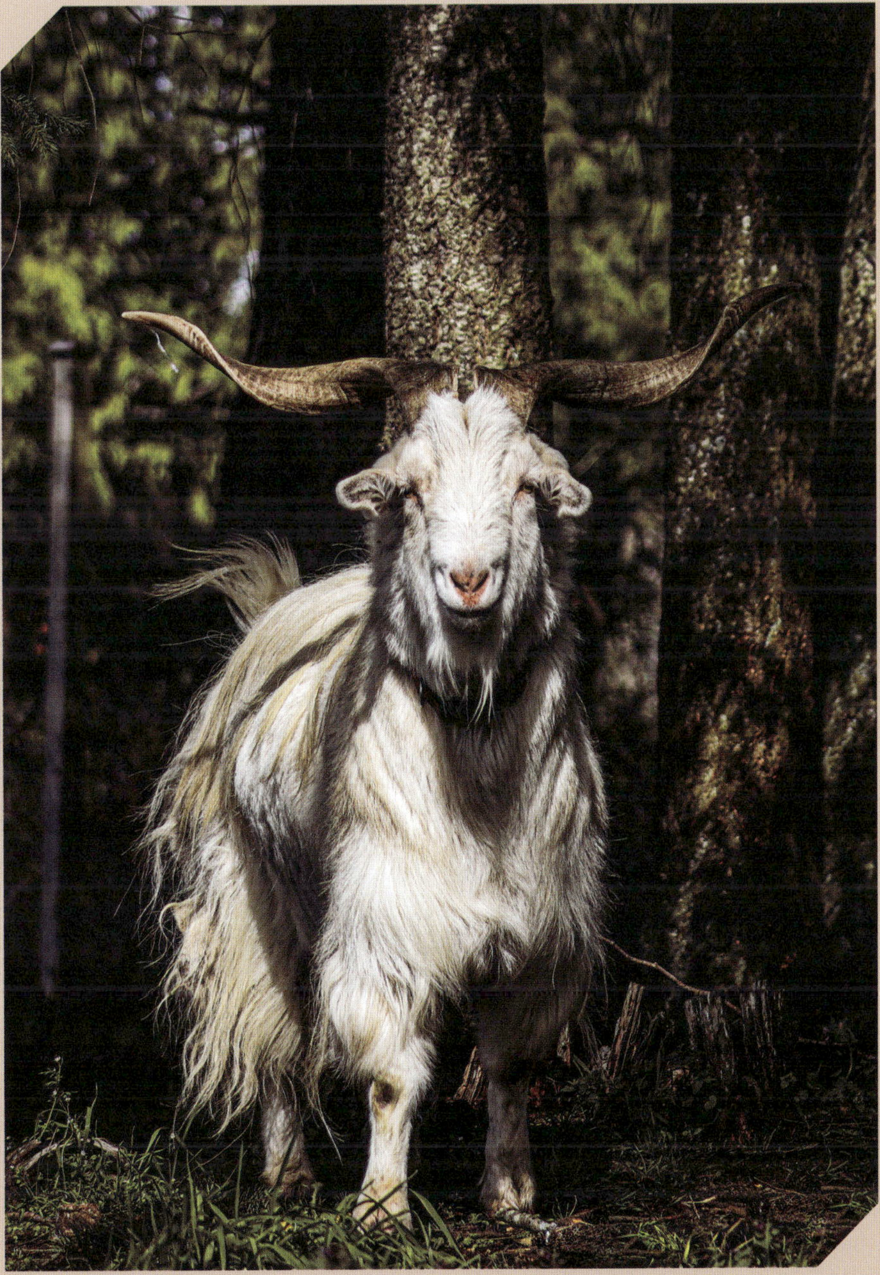

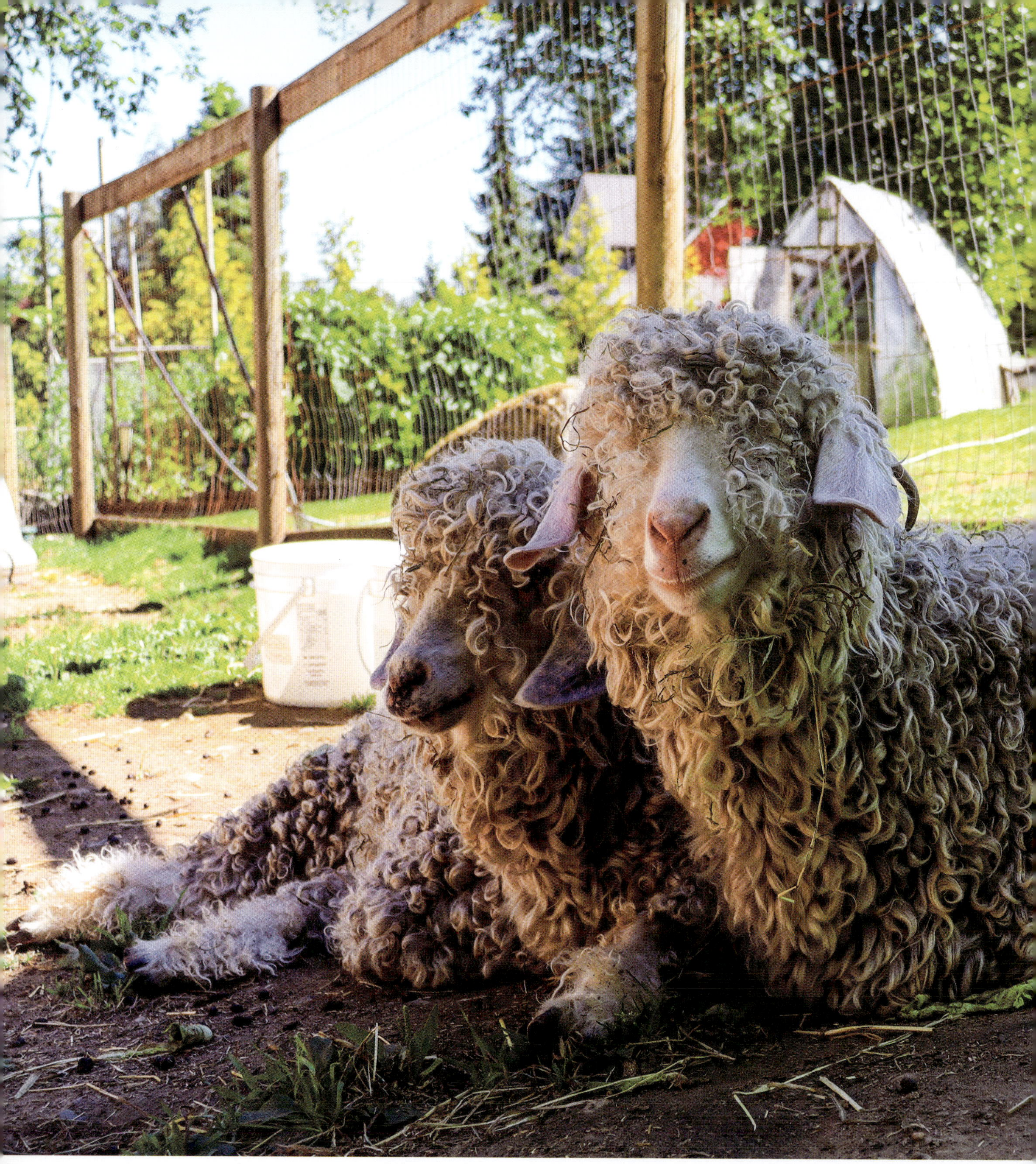

◀ Angora goats at Up A Creek Farm.

The farm used to be located on Salt Spring Island, and that's where Alice the Angora goat was brought into the herd. As a small kid, she was stranded on some rocks at high tide and passing boaters spotted her and came to her rescue. Salt Spring must have been the closest marina, so they brought the little kid there to the vet. The vet contacted Cara, thinking she would be a good fit for the kid. No one is 100 percent sure how the kid got on those rocks that day, but some speculate that she could have been from the feral goat herd on Saturna Island. At some point, the cashmere ram and the Angora goat got together and made a cashgora.

PRODUCTS: Raw fibre when available.

Mohair

Mohair refers to the curly, silky locks that grow off the Angora goat. The history of this ancient breed is an interesting one. Before the fifteenth century, the breed grew mainly in Ankara, Turkey, where it got its name. The Turkish clothes and threads produced from these animals were noticed and considered valuable, and the desire for the breed grew. For a time, the Turkish government resisted exporting the delicate breed, but eventually it was exported to France and the United Kingdom. By the nineteenth century, the breed had spread to much of the rest of the world.

Angora goats were first brought to Canada in the early 1920s from the US. As the pure breed was bred with other hardier breeds, it evolved to withstand our harsher climate, though the characteristics of the fibre stayed much the same.

The breed is small to medium sized. Both the males and females are adorned with horns, but the male horns are often spread out farther with a more defined curl, whereas the females tend to go straight back with less curl. Initially only white, the breed can now

▲ Angora kid under its mother at Up A Creek Farm.

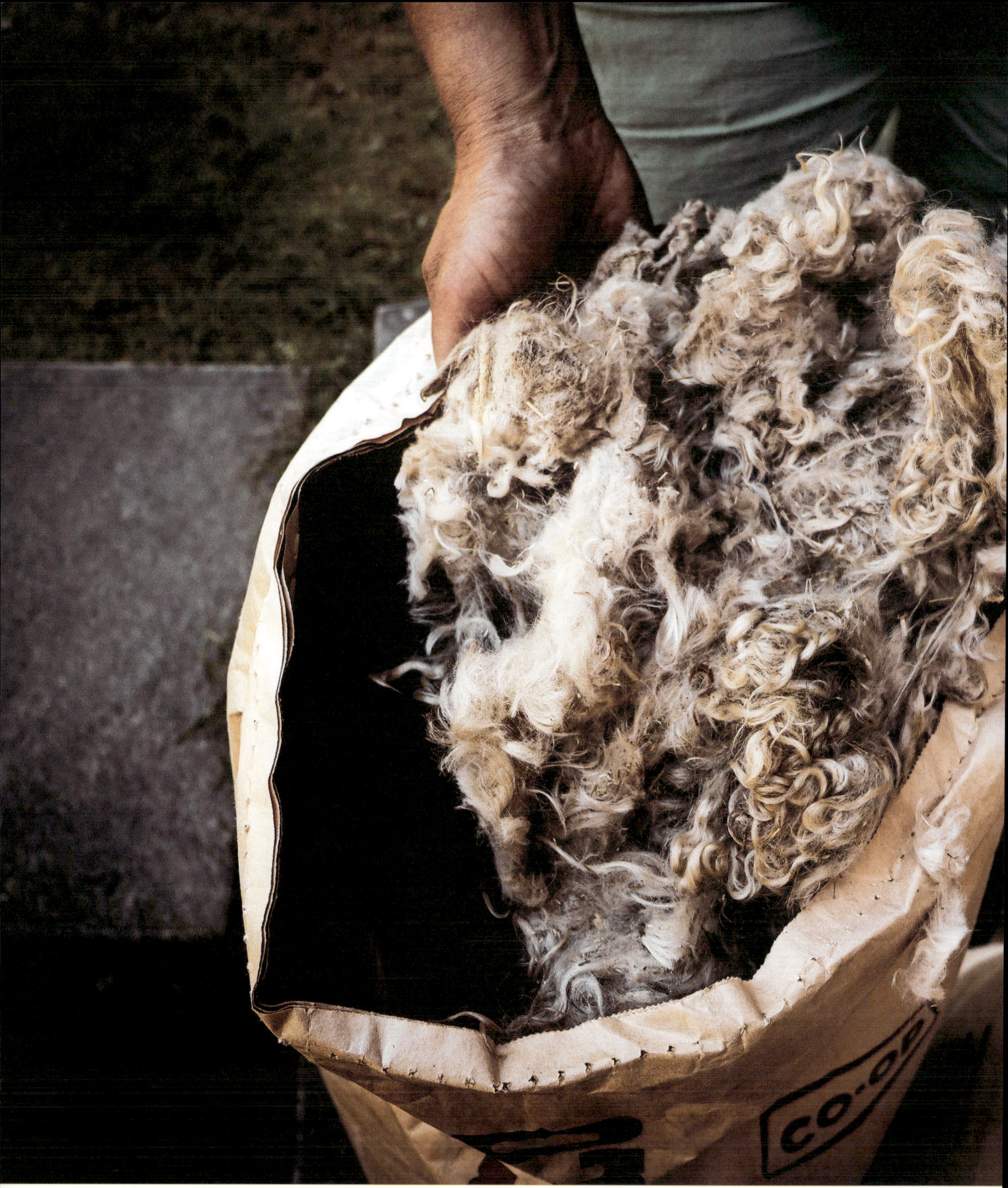
CO-OP

◀ Raw mohair fibre from Up A Creek Farm.

be found in a range of colours. However, groups like the Canadian Goat Society have breed standards that require mohair to be white, of a small diameter, not resembling sheep wool, and having a low percentage of kemp (coarse, straight fibre).

Up A Creek Farm & Fibre Mill, Qualicum

Pulling into this property, I can see why Jen and Mark Sterckx fell in love with the tall, three-storey home with its large wraparound porch. Tall trees shade the house and give the property privacy, but the yard is open and sunny, giving the flowering gardens the best space.

The Angora goats have their own fenced area to play and relax in, with outbuildings to hide in, on, and under. When we go into their area, they are calm and friendly and not jumpy around me. There are ten goats in the flock, including two small kids, one white male and a brown female that stay close to their mothers, nursing, jumping, and hiding under the bench in their shed. The one mother makes a throaty humming sound; Jen explains they only make this sound when they are pregnant or have a kid, as a way to communicate with their young.

The goats' fleece is extremely soft and grows very fast. The goats are sheared twice a year, a fall clipping around September and a spring clipping around March.

As we leave the goat area, Jenn points to a nice level area at the top of her property, near the road. She explains that she is in the process of getting the final funding together for building a fibre mill there. The property has already been approved by the district and plans are underway. Things take time but I can tell she is very excited and would love to have the mill running tomorrow. The hope is that the mill will give island farmers a more economical option for processing their fibre.

PRODUCTS: Raw mohair, yarn, washed and carded mohair.

PART II

Cellulose-Based Fibres

SO FAR, I have discussed the wide variety of protein fibres that the islands house. But how about the vast amount of greenery that surrounds us? What about all the sprawling farms? The islands grow a large amount of native and invasive plants that can be used as fibre. The number of plants being grown commercially for fibre, however, is small.

Cellulose-based fibres are harvested from the stems, leaves, or bolls of plants. The most common and readily found are cotton, hemp, and flax, which becomes linen.

Vancouver Island has the climate and agricultural resources to grow a variety of cellulose-based fibres for textile production. Hemp and flax have been grown agriculturally and experimentally on Vancouver Island for some time. What we lack is the infrastructure to process the plants into commercially saleable goods. Just like protein-based fibres, cellulose fibres take several steps to process into yarn or cloth, which then can be made into something people want to buy. People willing to do the work themselves have the opportunity in our climate to grow a backyard garden with a variety of fibre-producing plants.

Using cellulose-based fibres on Vancouver Island and the Gulf Islands is not new to the region. The Coast Salish Peoples worked with a variety of plants, both gathered and grown, to make all kinds of useful items. All parts of the cedar tree were used in clothing, transportation, dwellings, and functional items. Nettles, dandelion stems, milkweed fluff, bullrushes, cherry bark, spruce roots, willow, and hemp have all been used in making traditional Coast Salish clothing and everyday items.

It is no longer viable for everyone to gather from our depleting resources. Only those who have been taught to gather on their own traditional lands respectfully and sustainably should do so. Fortunately, a sustainable form of foraging and gathering has emerged. The removal and use of invasive species and green waste like English ivy, Himalayan blackberry, yellow flag iris, and Scotch broom is helpful and productive for our environment. These are a few of the intruder plants that have begun to take over our island's green landscapes. Organizations like EartHand Gleaners Society are working to educate and create community-based projects that lead to environmental awareness while modelling a method for building societies that nurture humanity.

Invasive plants can be removed and reused to create green, compostable art, structures, fences, garden-support systems, baskets, jewellery, and other useful items. For more on this topic, I recommend reading *Common Threads* by Sharon Kallis. She is the founder of the EartHand Gleaners Society and packed her book full of amazing information about eco-art and creating it sustainably.

In this section, I go into more detail about flax and hemp fibre. Unfortunately, I wasn't able to source anyone on the islands selling the raw materials, but I highlight the people and projects that are growing the viability of producing these fibres locally.

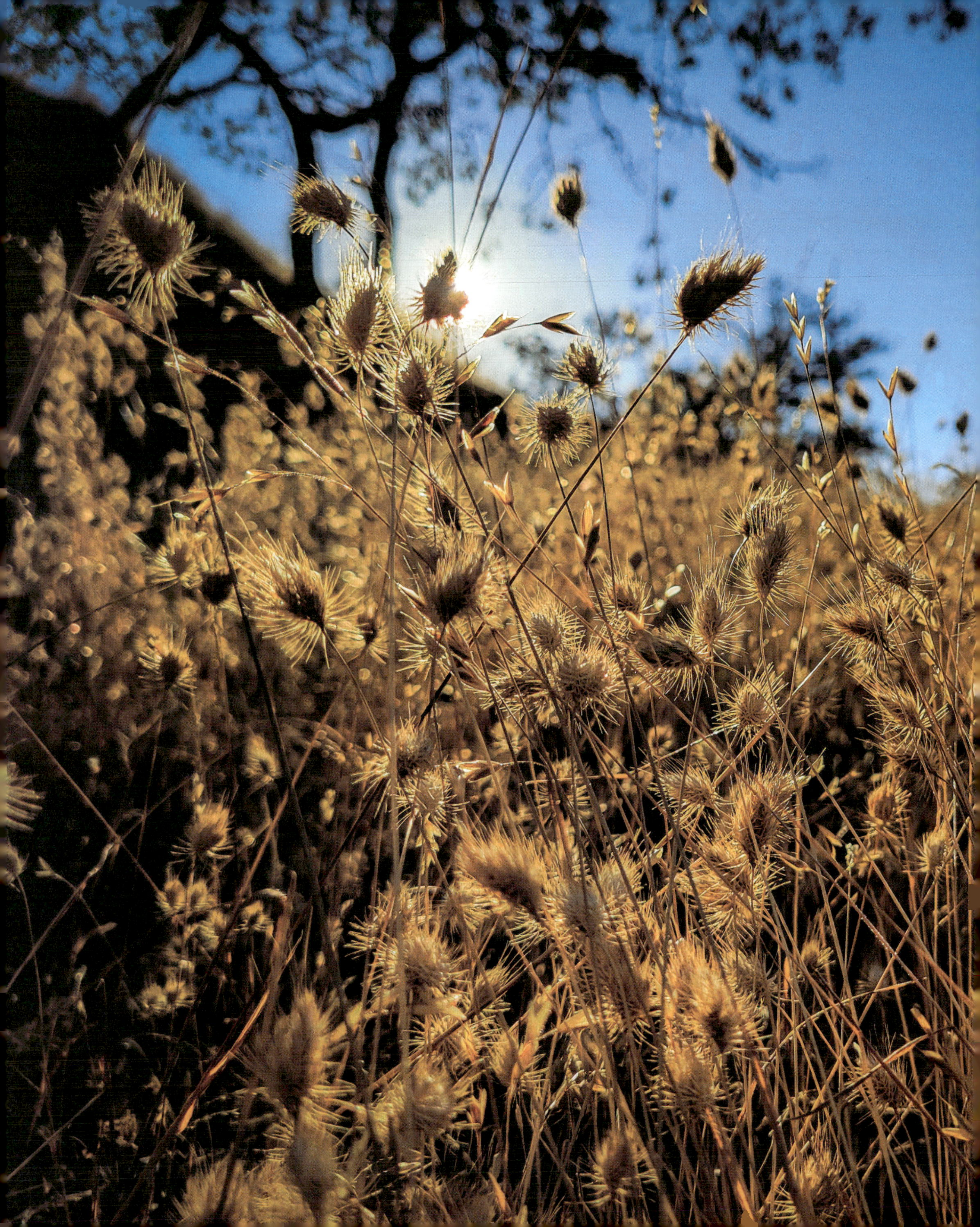

◀ Wild grass growing on Vancouver Island.

▲ English ivy weaving by Juliana Bedoya. JULIANA BEDOYA

▲ Himalayan blackberry fibre.
JULIANA BEDOYA

► Removing the thorns from Himalayan blackberry stems.

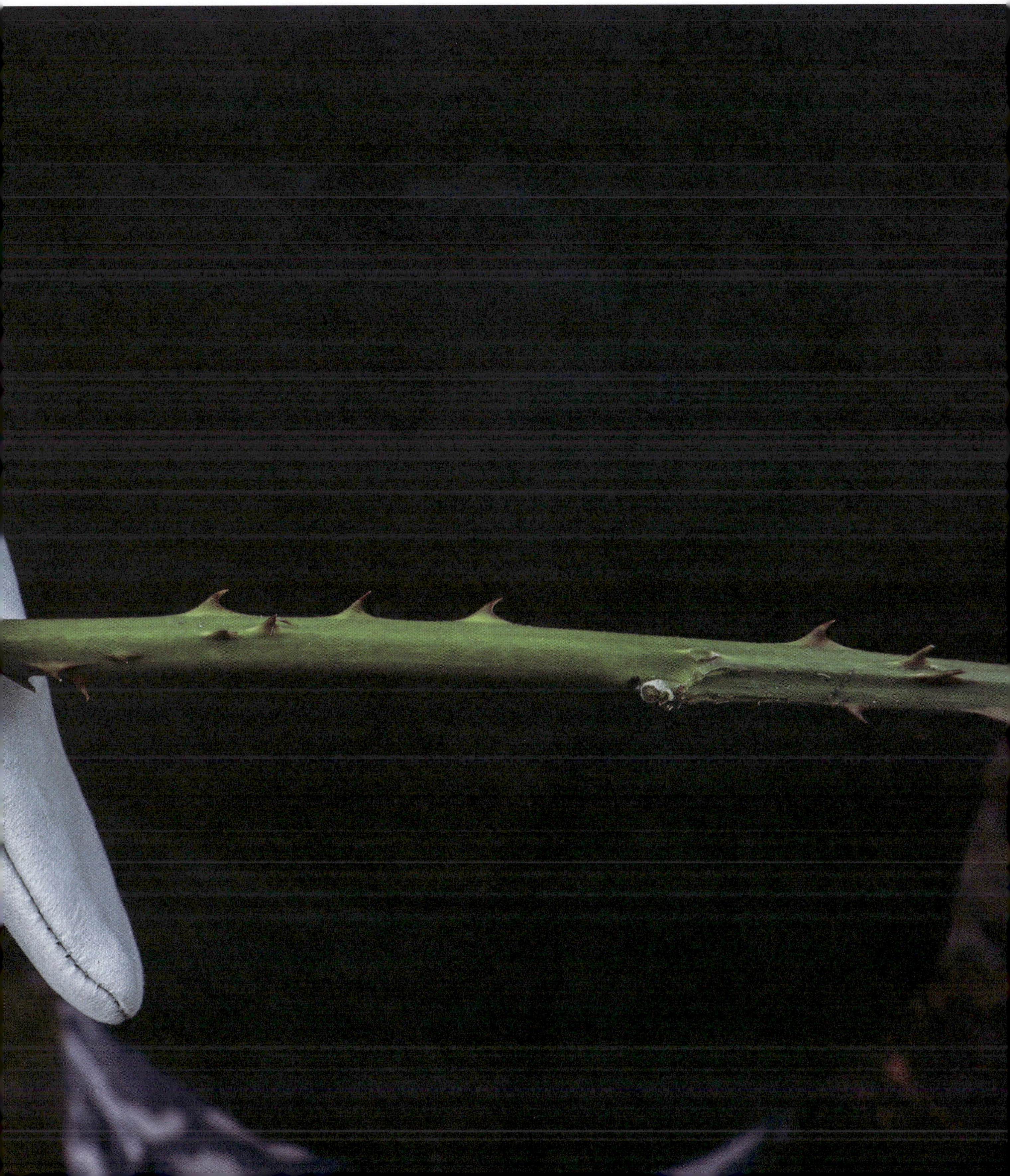

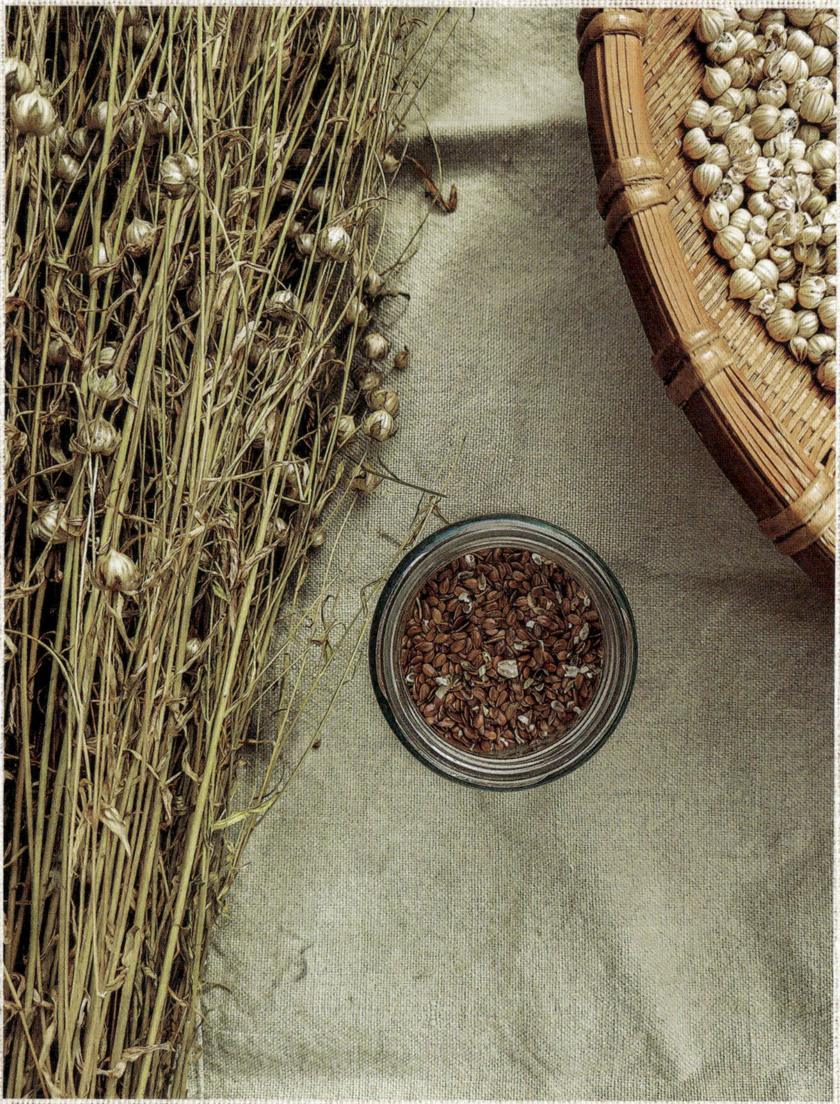

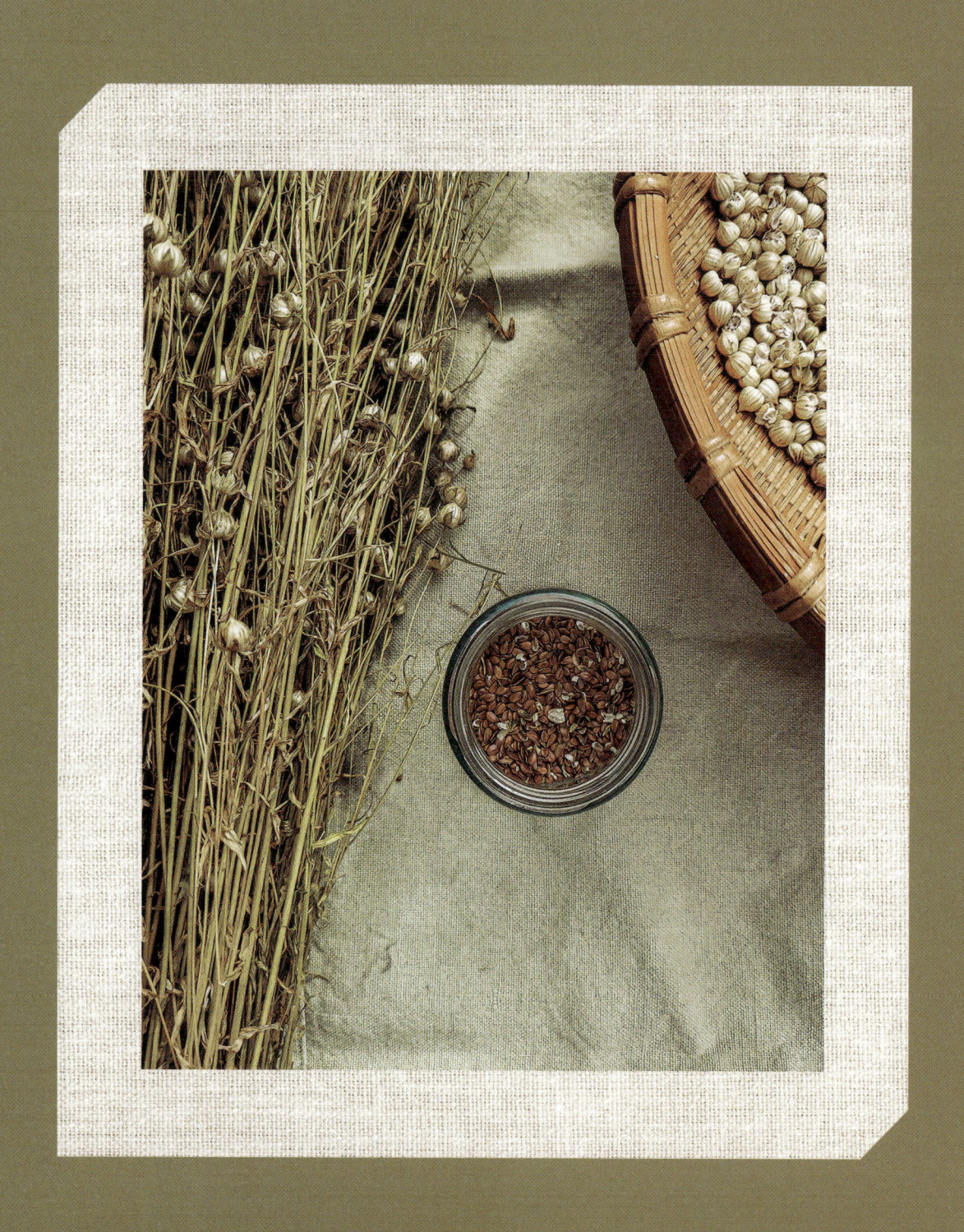

6 Flax

◀ Flax plants, seed heads, and removed seeds.

THE HISTORY OF turning flax (*Linum usitatissimum*) into linen is fascinating, spanning cultures and millennia, back to Paleolithic times. A group of archaeologists excavated Dzudzuana Cave in the eastern European country of Georgia and reported their findings in 2009. After digging up a twelve-foot layer of soil covering areas that held evidence of human occupation, they uncovered several flax fibres confirmed to date back 36,000 years. Some of these fibres were even dyed and spun or knotted, leaving archaeologists to conclude that they were used to make clothing and other useful items. People have depended on flax for a long time!

Domesticated crops of flax were first grown in North America in the nineteenth century, when colonists were attempting to curb the growing debt to Great Britain by boycotting the importation of fabrics. Many women had begun spinning and weaving their own cloth, not only for clothing and other apparel, but for household items like towels, bed linens, bow strings, cheesecloth, bread bags, money, sewing thread, and boat sails. Here in the Pacific Northwest, we began working with flax in the mid-nineteenth century.

Linen was popular until the infrastructure for processing cotton developed faster. In 1793, the cotton gin was invented, making cotton cheaper to produce, requiring less labour. By the 1800s, the only process that was mechanized for processing flax was the breaking of the stems. By 1825, a wet-spinning method was developed, but it required adept labourers to run the machines. Processing flax to

▶ Flax field in Saanichton.

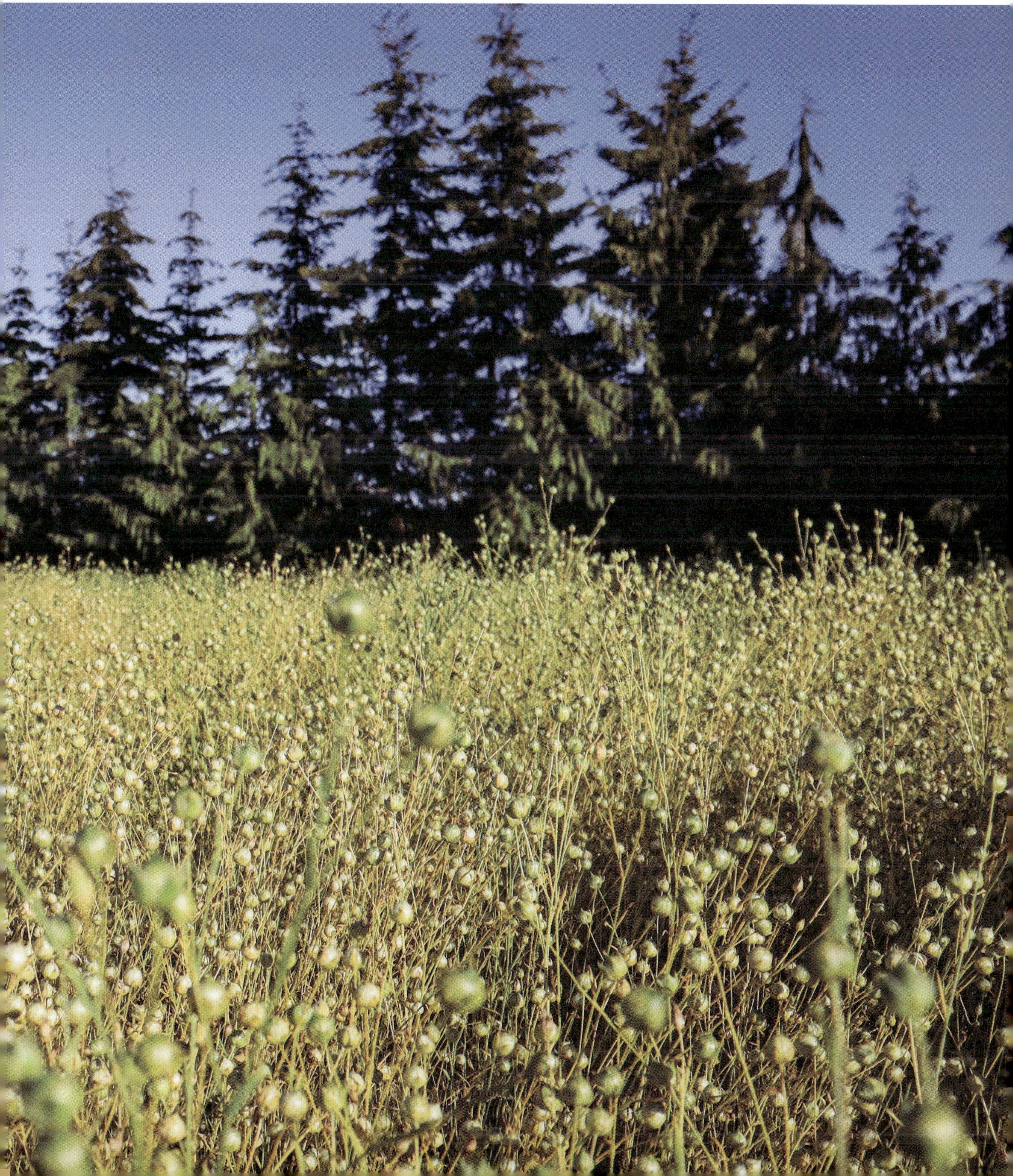

▶ The last of the flax flowers turning to seed in the one-acre flax field in Saanichton.

linen was a labour-intensive and highly skilled process that didn't suit mechanization, therefore the fibre was less desirable than the more easily processed cotton.

For a time, linen nearly vanished from our closets and homes. However, it's been making a comeback in recent years, with quite a few *cut and sew* businesses rising with the promise of ethically made linen clothing and apparel. Along with this resurgence of linen comes the demand for knowledge. Local guilds, fibre groups, and linen projects are popping up all over North America. These groups are working to clarify the process of growing flax and processing it into linen and to educate others on the possibilities.

Flax to Linen, Victoria

In 2010, a group of local fibre enthusiasts began an experimental growing project. With the help of a local farmer in Saanich who donated an acre of land, the group was able to plant the island's first large crop of flax specifically for processing into linen.

This kind of research project helps us understand what our region needs for building this textile fibre into our economy. It brought together farmers, spinners, dyers, weavers, designers, educators, and mechanics to streamline and clarify the process of turning flax to linen and then linen to cloth. The group also made and demonstrated pre-industrial tools at local farmers' markets and fairs.

These projects are about community. They are about going back to something that used to work for people in a regional way. Villages used to grow enough linen to make clothes, towels, and bedding, creating jobs and supplying people with textiles that were sustainable and local. Projects like this point out how far off track we have gone with our textiles.

Covid slowed the progress of the group, but they are currently back demonstrating the flax-to-linen process at markets and fairs. The group has new members who will hopefully take on leadership roles and continue its growth and progress.

One member of the group, Carol Hyland, planted a one-acre plot of "Lenore" flax in Saanichton in 2022. The planting worked as a perfect rotational cover crop for Field Five Farm pastures, where grain is grown for local distilleries.

After one hundred days of growing, Carol put a call out to the public to come and harvest some of the flax and learn about the planting process. Of course, I went. As we pulled flax on the last full moon of the summer, Carol explained that the flax worked exceptionally well as a cover crop because the dense planting of fifty pounds of seed per acre left little room for weeds to germinate.

Though this planting gave adequate coverage, the group is considering upping to sixty pounds of seed in future. Carol explained

▶ A demonstration at Fibrations by a Flax to Linen Victoria member, showing the break tool (top), scrutching (middle), and hackling (bottom).

◀ Flax being spun into linen by a member of the Flax to Linen group.

that the more densely the flax is planted, the taller the stems will grow before branching out. This gives longer fibres in the end and leaves the farmer with a field free of weeds, ready to plant again once the flax is harvested.

Linen has many desirable qualities when it comes to wearability. Garments are durable and linen is cool to the touch, making it a wonderful warm-weather fabric and a lovely choice for bedding. The more you wear and wash linen, the softer it becomes. Linen can hold 20 percent moisture before it begins to feel damp, and it is quick drying. Garments made with linen will shrink less than cotton and hold their structure better. As well, linen not processed with chemicals is biodegradable, going 100 percent back into the earth to regenerate our soil.

For more detailed resources for growing and processing flax, I recommend two books. The first is by Raven Ranson, called *Homegrown Linen: Transforming flaxseed into fibre*. Raven lives on Vancouver Island and works with the local Flax to Linen group. She has done extensive research on growing and processing flax on her farm. The second book is by Linda Heinrich, called *Linen: From Flax Seed to Woven Cloth*.

PRODUCTS: Educational website and a network of flax processors.

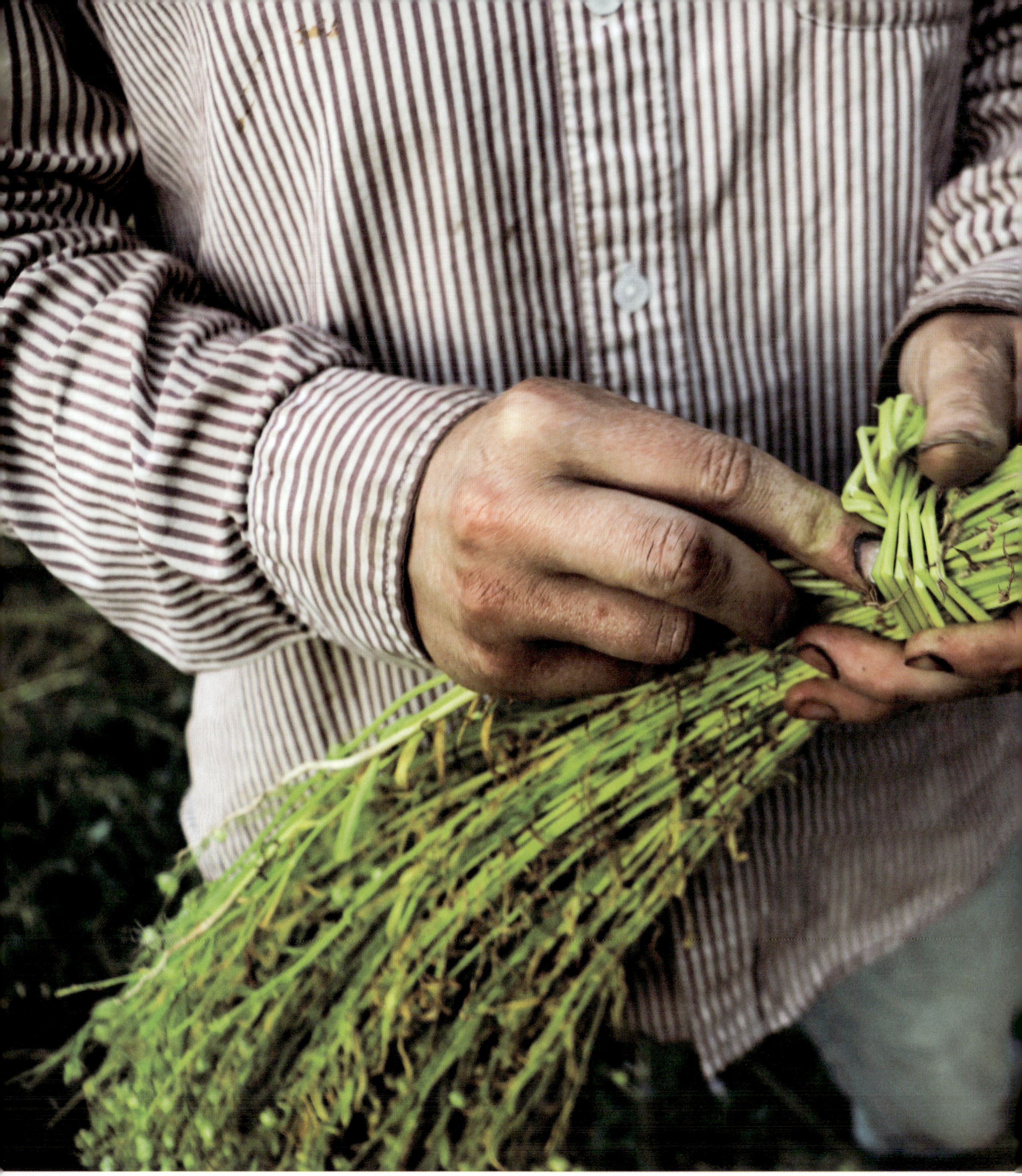

◀ A bundle of freshly harvested "Lenore" flax.

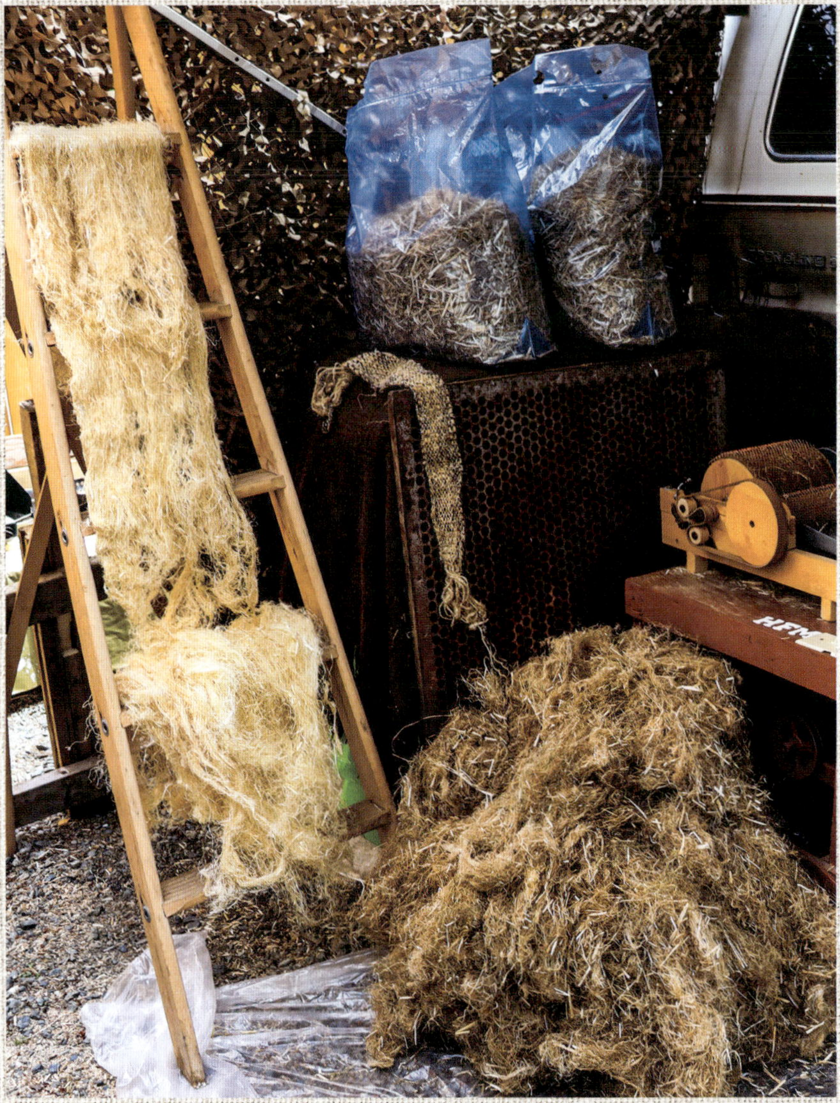

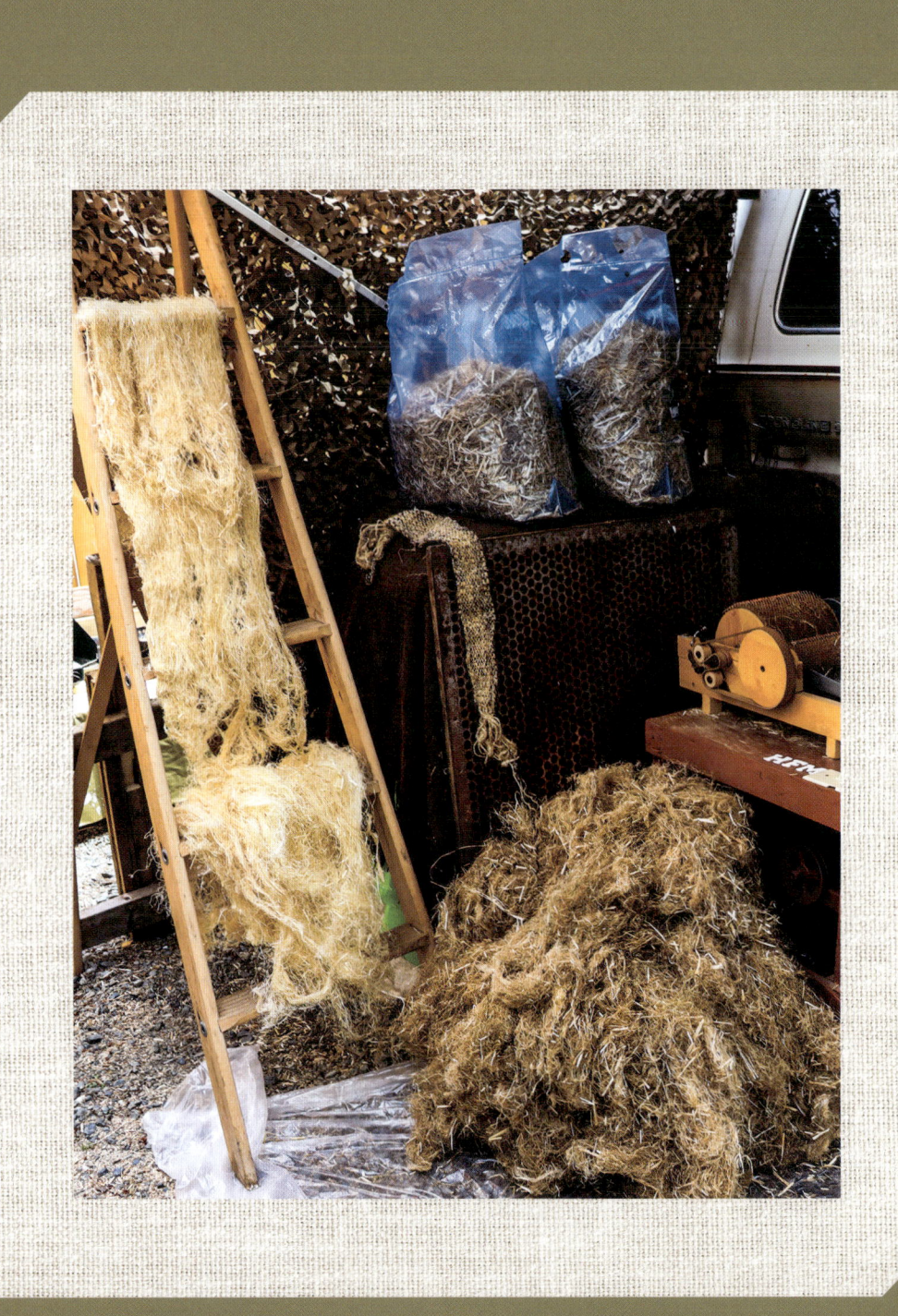

7 Hemp

◀ Heritage Fibre Mill's hemp processing station.

HEMP WAS a popular fibre in North America until cultivation of the plant was banned in the US in 1937 and Canada in 1938. The hemp industry collapsed. Hemp got classified as illegal along with marijuana. There is speculation among hemp growers that politics played into the demise of the hemp industry.

Hemp is a quick growing (85 to 120 days), renewable resource once used in several industrial applications, including paper. In 1916, a study by the US Department of Agriculture reported that 1 acre of hemp equals 4.1 acres of trees. The study predicted that by the 1940s all paper would come from hemp. However, hemp was still banned and wasn't grown agriculturally in Canada until government research done between 1994 and 1998 showed that hemp could be grown separately from cannabis, and laws were amended to allow for the growth of hemp again. Under the Cannabis Act, regulated by Health Canada, people must apply for an industrial hemp license to possess, produce, and distribute any hemp or hemp products legally.

Given this long, complicated history with legally cultivating hemp, farmers and those looking to start a new agricultural endeavour might not want to jump through the hoops to start this crop commercially. A few farms on Vancouver Island have attempted to grow hemp for agricultural uses, but people are still leery of its relation to cannabis, and farmers have had their crops stolen by people thinking it was marijuana.

▲ Hemp products from Heritage Fibre Mill.

▶ Woven Hemp at Heritage Fibre Mill.

I wasn't able to locate anyone currently growing hemp agriculturally on the island. However, there is someone who is exploring future possibilities for hemp in our region.

Heritage Fibre Mill, Qualicum Beach

Imagine a mobile hand mill that teaches people the multi-step process of turning plants into fibre. That is exactly what Heritage Fibre Mill (HFM) is endeavouring to build. Terence Bexson started HFM so he can demonstrate and teach people what it takes to create bast fibres. The mission is to create a renewable product with as small a footprint as possible.

HFM is currently working with a range of fibres, but focuses mainly on hemp, flax, and nettle processing. Terence has acquired, made, and sourced all the tools needed to process and make these fibres into usable materials. The local spinning and weaving community has been very supportive, donating tools, wool, and resources. He works with a gorgeous antique hackle, a donated drum carder, and multiple spinning wheels to get desired results.

His workspace is condensed but well organized, with an area for each step of the process. Just from visiting his space, I can visualize the steps it takes to make plant-based fibres. This labour-intensive process takes time and care, but the result is a sturdy, long-lasting fibre that is environmentally friendly.

Terence shows me the hemp in its different stages and how he makes a mixture of the "waste" fibre and lime to make hempcrete, which can be used to make a variety of things, most often alternative building materials for walls, subfloors, and roofs. Seeing this process proves that every part of the plant can be used.

Terence shows me a bracelet he made with a local youth group. He's been wearing the simple, thigh-spun loop of hemp for months, and the fibre is soft but still strong.

I run my hands along all the handwoven scarves and can't believe how soft the blended hemp and linen feel. They add structure and texture but don't compromise comfort.

The HFM website offers an interesting opportunity to handspinners with its Fibre Connection page, where makers can sell local handspun fibre.

PRODUCTS: Hemp in different stages, hackled flax fibre, wool, handwoven goods, and shea butter soap with hemp.

▶ The making of hempcrete at Heritage Fibre Mill.

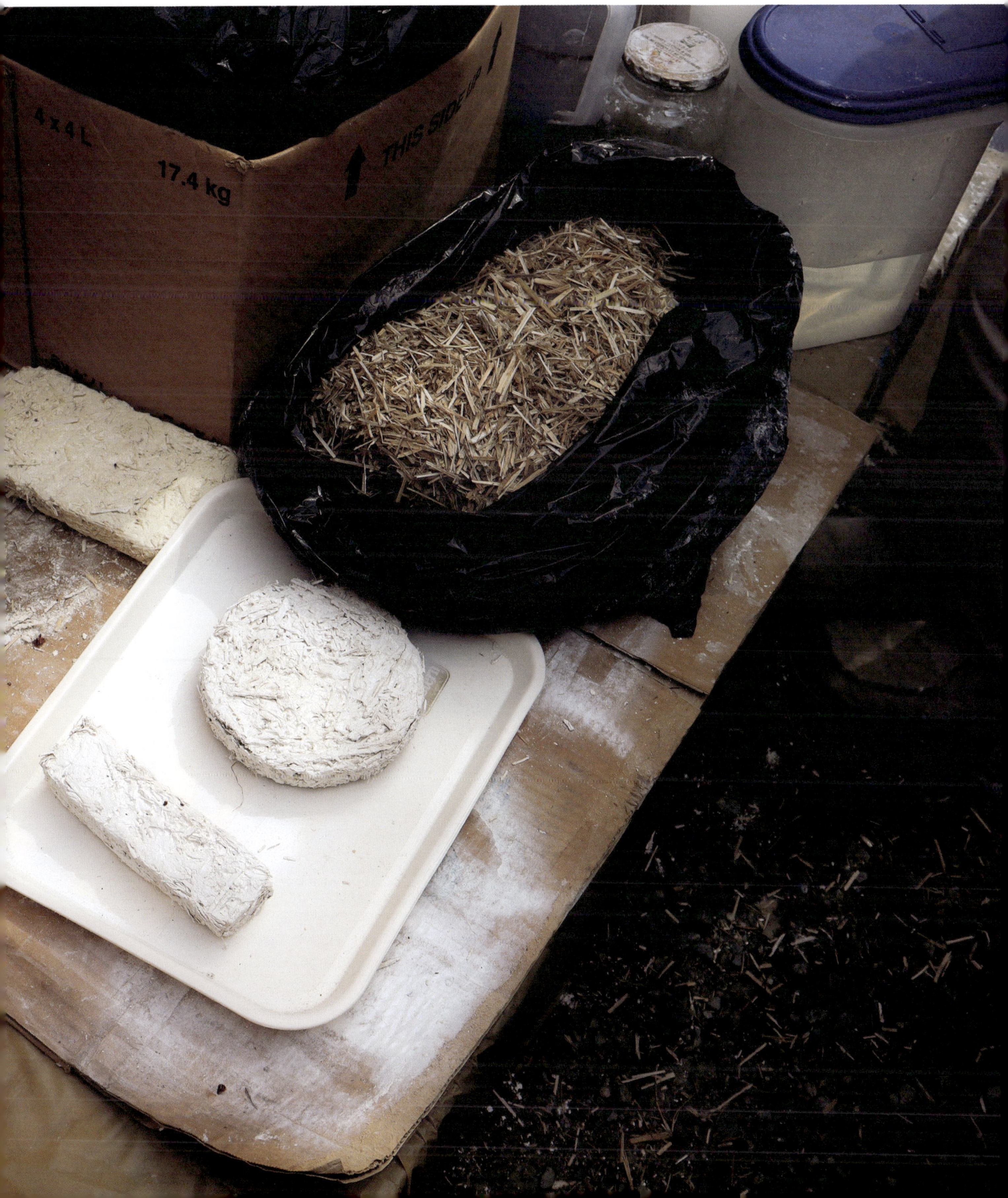
4 x 4 L
17.4 kg
THIS SIDE UP

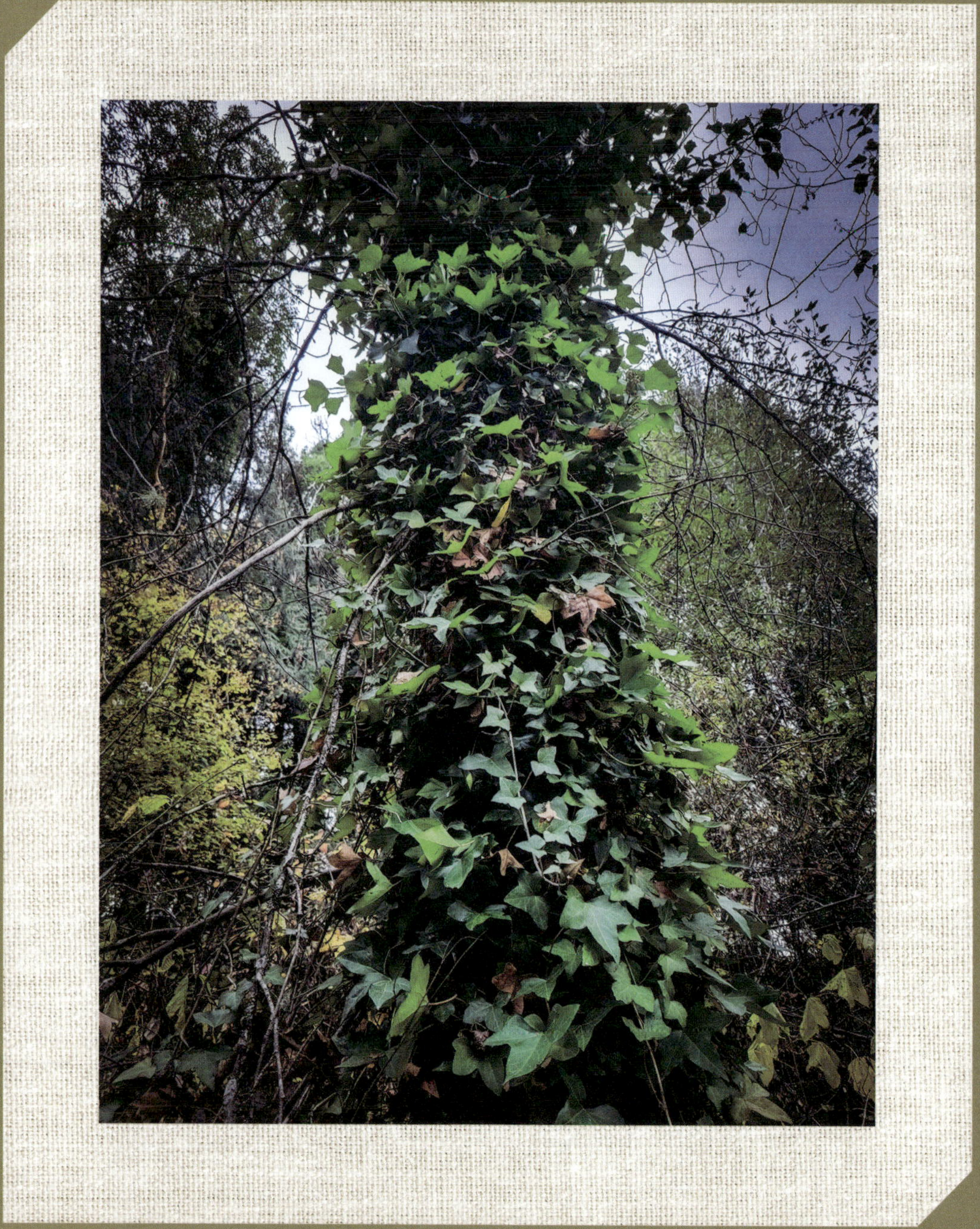

8 Invasive Species

◄ English ivy overtaking a tree at Transfer Beach in Ladysmith.

VANCOUVER ISLAND and the Gulf Islands are brimming with invasive plant life. These non-native plants are persistent and are doing damage to our local biodiversity with their dominant ecological destruction. According to the Capital Regional District, we have "potentially the highest diversity of invasive species in the province."

Because these species are not native to our climate, they are often unaffected by our weather and have no natural predators, which helps them thrive. If we leave these invasive species to grow, they will rapidly dominate our landscapes and leave little room for native flora and fauna, leading to a dramatic loss in diversity.

Many of these invasive species have been introduced to our landscape through the spread of seeds by wildlife, illegal dumping of garden waste, movement of soil, and unknowing gardeners thinking they've found a fun new ornamental plant. A few of the most common are Scotch broom (*Cytisus scoparius*), Himalayan blackberry (*Rubus armeniacus*), English ivy (*Hedera helix*), morning glory (*Convolvulus arvensis*), and yellow flag iris (*Iris pseudacorus*).

One way we can use some of these plants is to process them into bast fibres and then into usable items. Sharon Kallis puts it perfectly in her book *Common Threads*: "We should work not just with what is at hand in our immediate environment, but with what is abundantly at hand and needs to be kept in check."

This reciprocal relationship with the land makes working with these bast fibres a revolutionary act. It brings back practices that

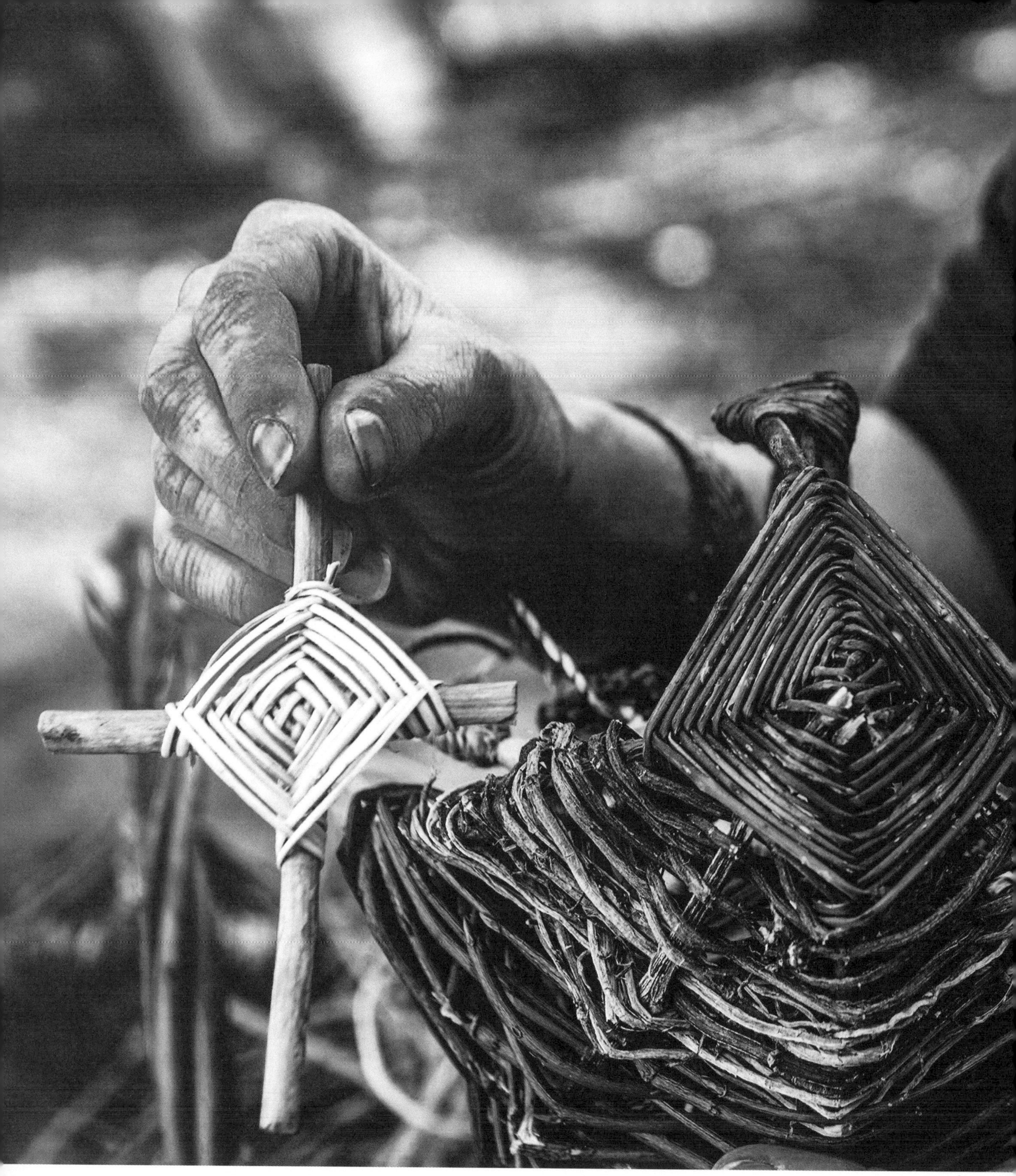

◀ A handwoven English ivy basket made at a Removal of Invasives for Basketmaking class taught by Plants Are Teachers. STEPHEN HAWKINS

▲ Himalayan blackberries at the end of the season.

▶ Morning glory taking over lavender.

Opposite Himalayan blackberry thorns.

foster the native ecology and values the land over the material. Not all of our materials need to be "beautiful" to start. It's the work that's done by hand and heart that makes a handmade textile special.

However, when you go to harvest these plants as fibre materials, it's important to know what you are doing. Removing all of a plant, depending on the time of year and the landscape, may not be the best method. Some plants like Scotch broom should only be pulled while they are blooming; that is when the plant's energy is above ground, working on flowering, leaving it vulnerable. If you cut it while it isn't blooming, the energy below ground in the roots will work on spreading and growing another plant. I encourage you to research each specific plant before removing, harvesting, or processing it.

Plants Are Teachers, Comox Valley

If you are looking for a hands-on opportunity to work with local bast fibre, Juliana Bedoya teaches workshops through her company, Plants Are Teachers, in the Comox Valley. She is a spinner, weaver, teacher, and environmental artist who fosters working in the community.

▶ Small handwoven dandelion baskets made in a class taught by Plants Are Teachers. JULIANA BEDOYA

Juliana demonstrates how to ethically gather bast fibres from our region, both grown in the garden and foraged. The foraged plants are invasive species like stinging nettle, Scotch broom, morning glory, and Himalayan blackberry. Juliana teaches how to process and use these fibres, while also helping with the ecological restoration of our land by removing the plants.

Juliana's craftsmanship can be seen in each strand of cordage she spins. The baskets and art installations created from bast materials are sculptural, intricate, and beautiful to look at. She has a line of jewellery and wall hangings that incorporate river rocks, smoke-fired clay beads, birch bark, linen, and other bast fibres. Each piece has a clean design that complements the natural characteristics of the materials.

Juliana's work cultivates conversation around our connections with the land and the viability of running a business based on a symbiotic relationship with the environment.

PRODUCTS: Workshops in bast fibre and natural ink-making, handcrafted jewellery.

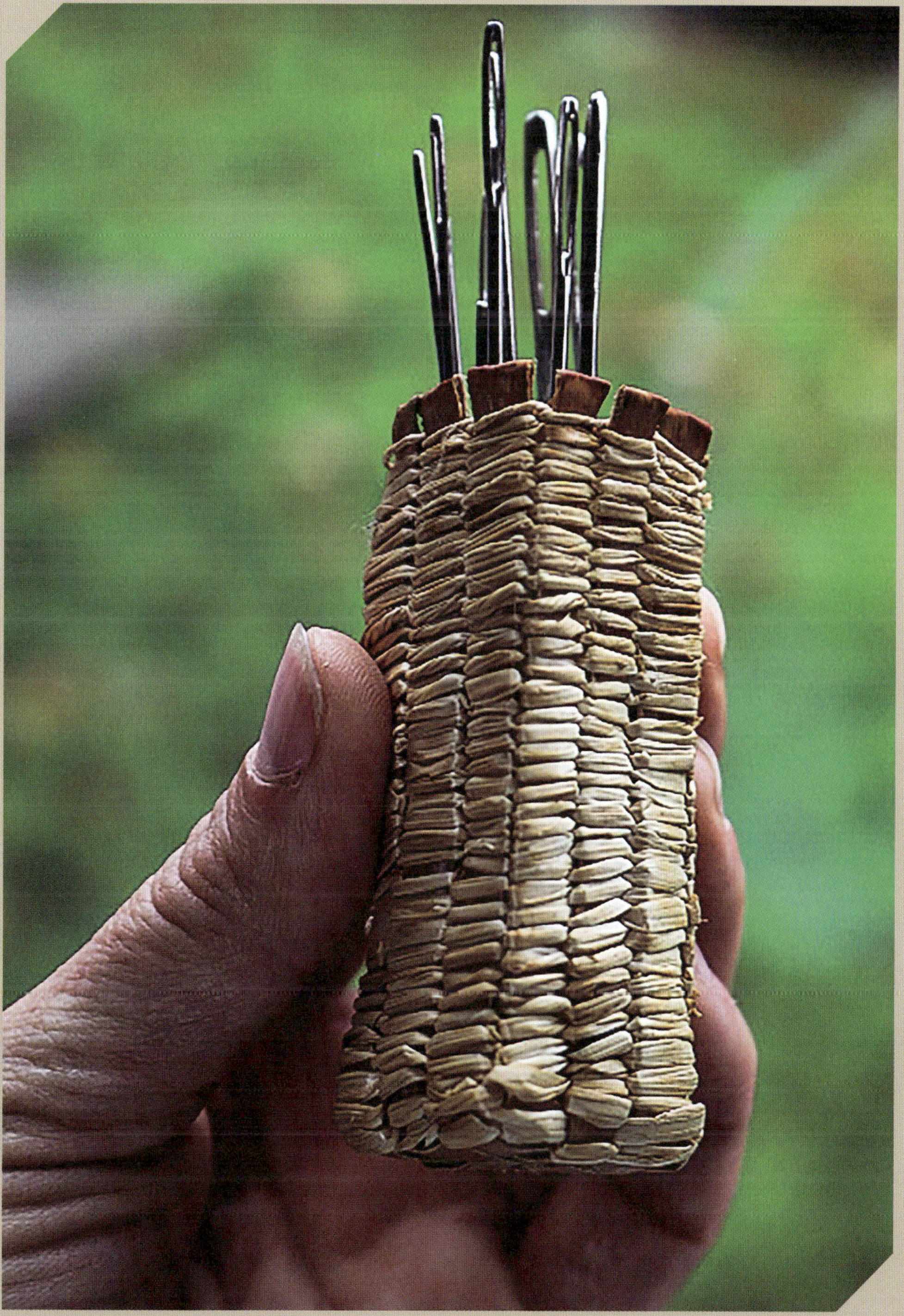

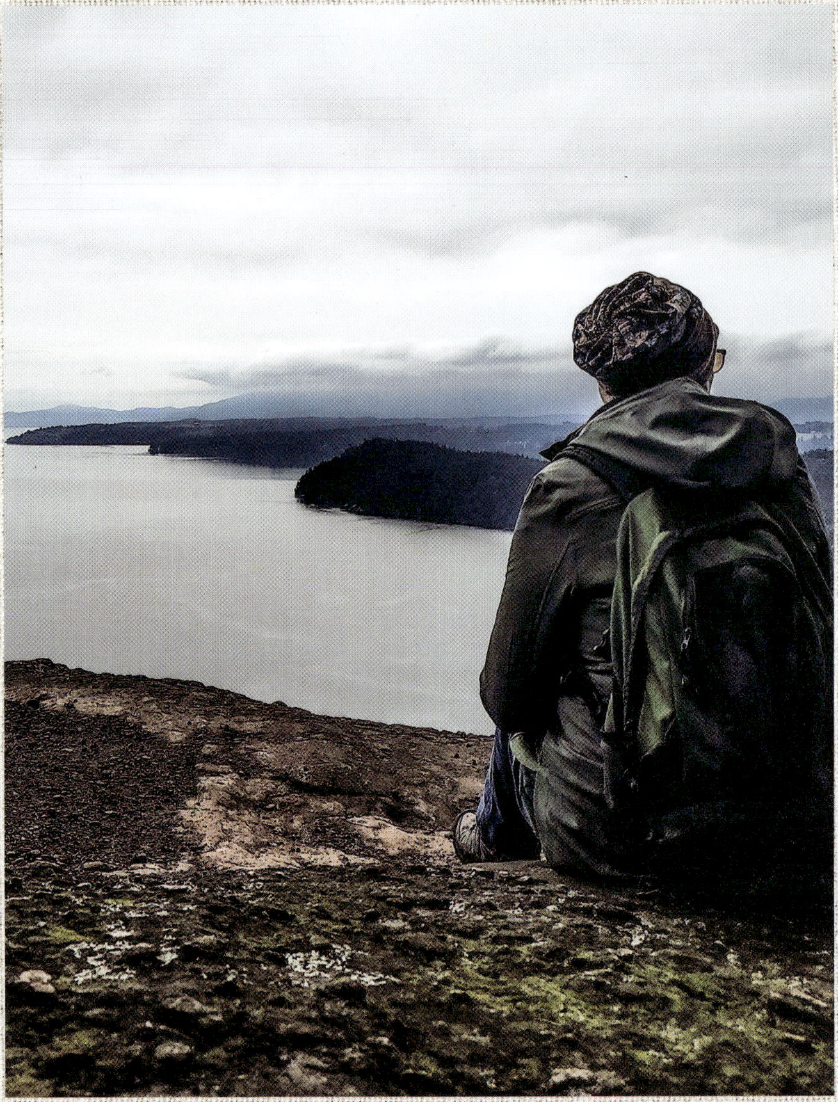

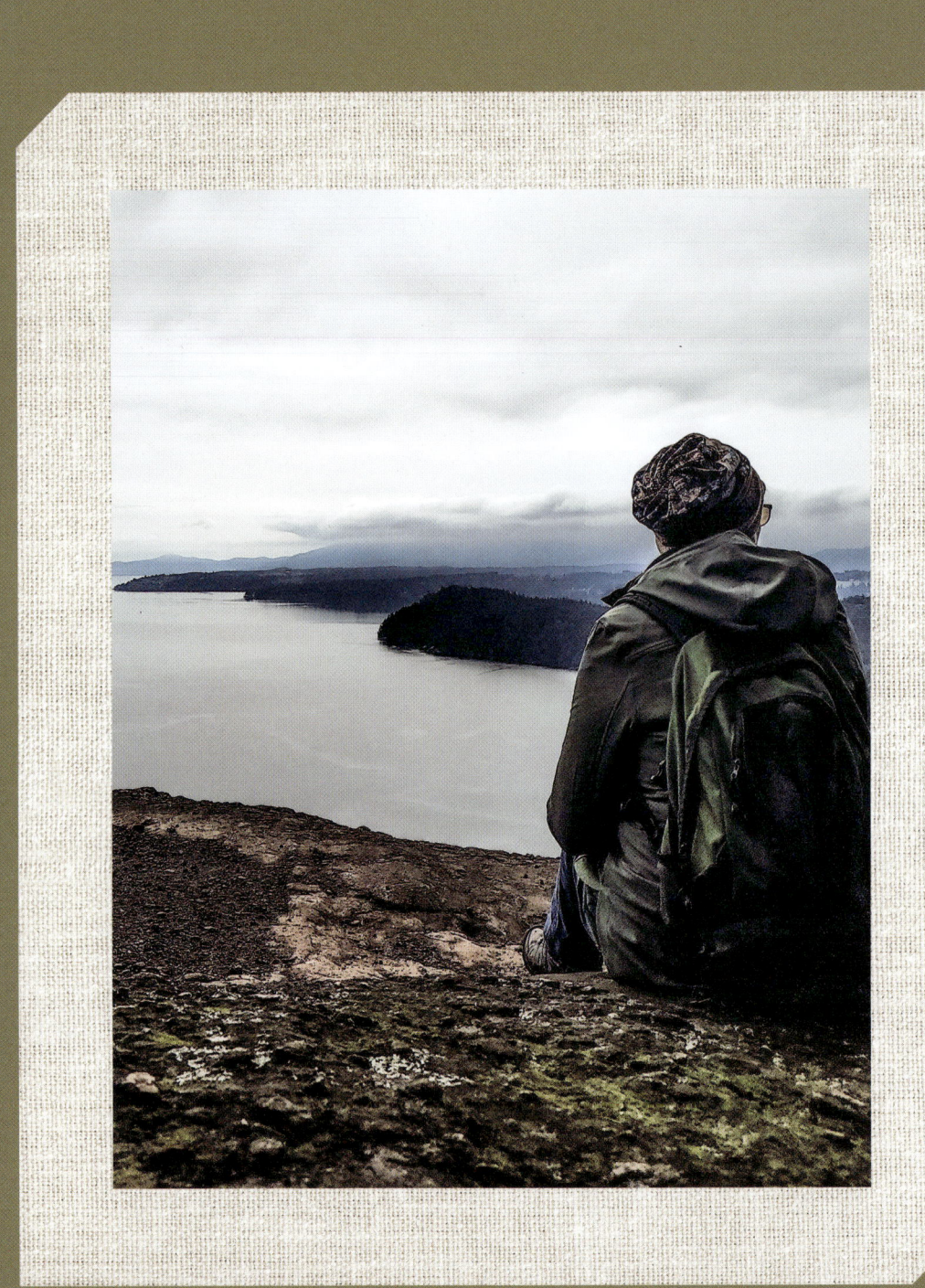

Conclusion

◀ Looking out over the Gulf Islands.

WHEN WE LIVE somewhere for a long time, we begin to take its beauty for granted. We stop noticing how the light catches and illuminates the individual strands of beard moss hanging from the maples. We forget that, watching the ocean at just the right moment, we could see a whale surface. We stop appreciating how the seasons take their time to change, and how precious are the things that come from the soil right under our feet.

Researching for this book reinstilled the unique beauty of this place. I got to explore Gulf Islands I hadn't been to, and the magic of this land seeped back into my bones. I remembered what brought me here and kept me here. The land and people of this region are magnetizing, and I was truly humbled by working with them.

As I conclude, I feel like there is so much more to say about the importance of supporting local farmers and a circular economy, and of being conscious of our purchases. Reconnecting with the land for our fibres and garments seems like the next logical step toward bringing our agriculture home.

After visiting these farmers, I can see how much dedication it takes to raise animals and use their fibre in an economy that doesn't support it.

Growth is happening in our region from within. The farmers are making it happen. With the hope of future mills, an operating pelletizer, and creative ways of using waste wool, farmers are already doing all they can to work in a circular fashion. Now they need our support.

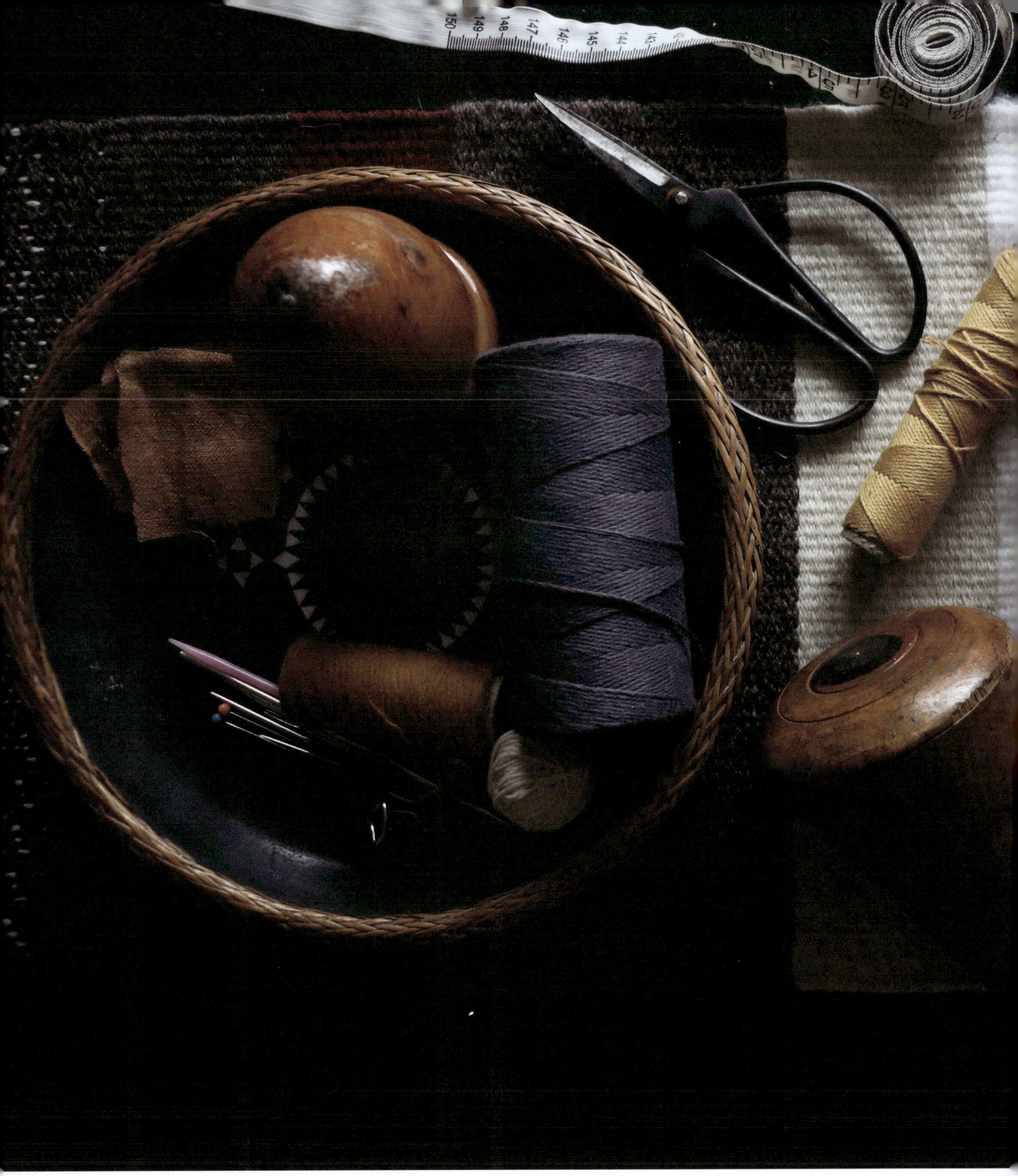
150 149 148 147 146 145 144 143

◂ Garment mending.

Each one of us can easily contribute by purchasing local fibre or fibre products, supporting farm stores that sell fibre products, and shopping at and donating old garments to thrift stores, many of which are charity run, so you're helping in two ways. You can pick up some knitting needles and learn to knit, take a sewing class, learn to mend, repeat, and re-wear old garments or rework them into something new. There are so many ways to keep garments circulating longer.

My hope is that you feel motivated to learn more about our current textile economy. The next pages are full of local resources I encourage you to explore. Join a fibre group, go to a fibre event, visit a farm, and stay fibre curious.

▲ Indigo plants growing in the garden at Hinterland Yarns.

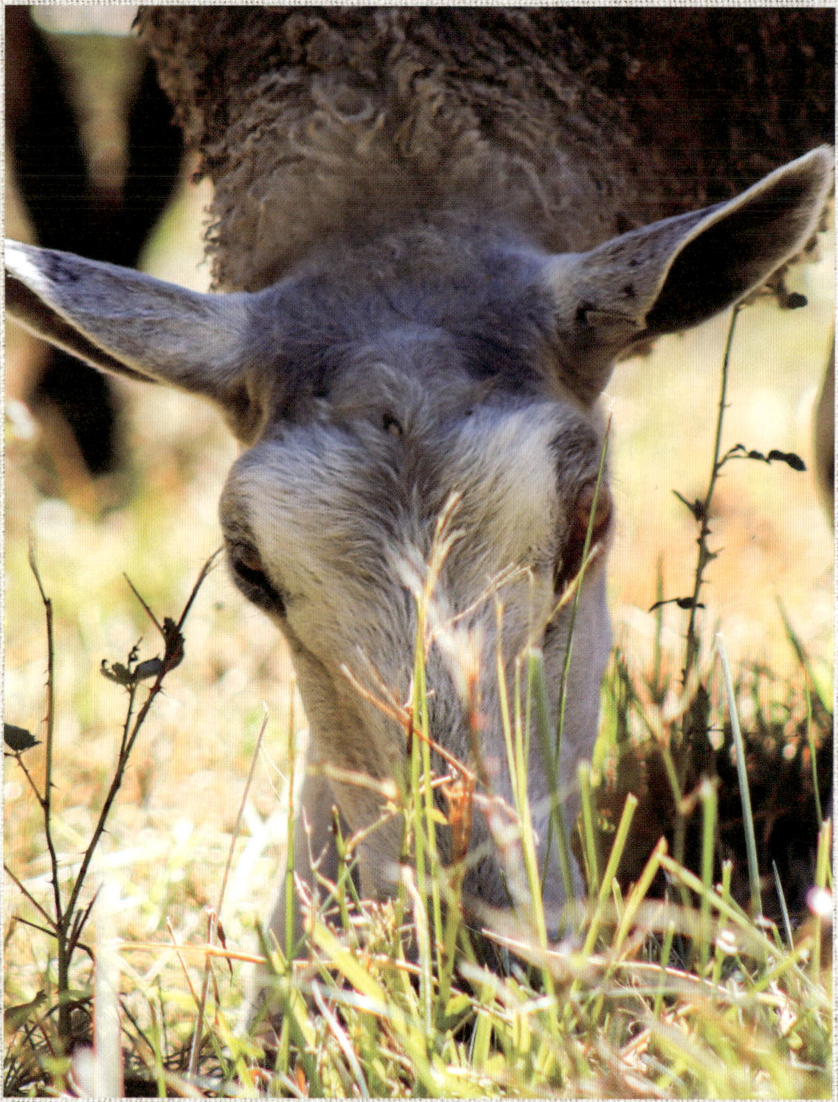

Resources

Local Fibresheds

Vancouver Island Fibreshed
Registered producers, local events.
- vancouverislandfibreshed.ca

Salt Spring Island Fibreshed
SAORI Salt Spring Weaving Studio.
- saltspringweaving.ca

Canadian Fibreshed Network
Cross-Canada network.
- canadianfibreshed.org

Guilds / Clubs / Associations

Victoria Handweavers & Spinners Guild
Community fibre events, workshops, programs.
- vhwsg.ca

Jane Stafford Textiles Online Guild
Online guild, weaving supplies, yarn.
- janestaffordtextiles.com

◀ A Bluefaced Leicester at Elf Leatherworks.

Qualicum Weavers and Spinners Guild
Workshops, study groups, mentoring.
- qualicumweaversandspinners.blogspot.com

Association of Northwest Weavers' Guilds
Classifieds, events, education.
- Northwestweavers.org

Salt Spring Weavers and Spinners Guild
- saltspringweaversandspinners.com

Gabriola Weavers and Spinners
- northwestweavers.org/gabriola-weavers-spinners

Tzouhalem Spinners and Weavers Guild (Cowichan Valley)
- tzouhalemspinnersweaversguild.com

The Bradley Centre (Coombs)
Bradley Thursday Spinners, Oceanside Fibre Connection, Tapestry Weaving, Oceanside Felters.
- thebradleycentre.com

Deep Cove Weavers and Spinners Guild
- dcwsweavers.blogspot.com

Denman Island Weavers and Spinners
- northwestweavers.org

Mid-Island Weavers and Spinners Guild (Nanaimo)
- guildmiws.blogspot.com

Midnight Shuttles Spinners and Weavers Guild (Campbell River)
- northwestweavers.org

Nanoose Bay Weavers and Spinners Guild
- northwestweavers.org

The Woolgatherers (Comox Valley)
- northwestweavers.org

Canadian Co-operative Wool Growers Limited
Education, events, wool buying. Serving Canada's sheep producers with pride since 1918.
- wool.ca

Vancouver Island Llama & Alpaca Club
Registered llama and alpaca breeders.
- vilac.org

BC Sheep Federation
Registered sheep breeders, shearing times, products.
- bcsheepfed.com

Canadian Cashmere Producers Association
Registered breeders in Canada.
- canadiancashmere.ca

The Canadian Goat Society
Goat breed information and statistics.
- goats.ca

Inter Island Sheep Breeders Association
- interislandsheepbreeders.ca

Stores / Online Shops / Educational Resources

Knotty by Nature (Victoria and Online)
Fibre, fibre processing supplies, fibre art supplies, classes.
- kbnfibres.ca

Chaotic Fibres (Brentwood Bay)
Quality wool, silk, and specialty fibres for spinners, felters, and knitters. Online and in-person shopping by appointment.
- chaoticfibres.com

Jill's Fibres (Online)
Local fibres locally processed.
- jillsfibres.ca

SAORI Salt Spring
Classes, workshops, retreats, material, equipment.
- saltspringweaving.ca

Maiwa School of Textiles (Vancouver and Online)
Online courses, information library, natural dyes and supplies.
- maiwa.com

EartHand Gleaners Society (Vancouver and Online)
Art, ecology, people.
- earthand.com

The Silk Weaving Studio

Silk spinning and weaving lessons.

- silkweavingstudio.com

The Small Bird Workshop (Online/In-person)

Classes and events, yarn, fibre, the Fibreworks Podcast.

- smallbirdworkshop.com

The Spool Sewing Studio (Courtenay)

Community sewing space, curated fabrics, patterns, kits, and classes.

- thespoolsewingstudio.com

The Spool Yard (Duncan)

Open sewing studio and sewing classes.

thespoolyard.ca

Woven: Community + Textiles (Cumberland)

Weaving classes.

- wovencommunityandtextiles.ca

Fern + Roe (Saanich)

Hide tanning workshops and tools.

- fernandroe.com

Vancouver Island Wool Pellets

100 percent island-produced wool pellets.

- viwoolpellets.ca

Canadian Wool

History, processing information, buyers guide.

- canadianwool.org

Knit Social (Vancouver and Online)

Community fibre events, webinars.

- knitsocial.ca

Our Social Fabric (Vancouver and Online)

Non-profit fabric store selling deadstock fabric and fibre arts supplies. They keep "waste" fabric out of the landfills.

- oursocialfabric.ca

Fabcycle (Online)

Selling deadstock fabric (what apparel manufacturers cannot use) to divert it from the landfill.

- fabcycle.shop

Young Agrarians

Farmer-to-farmer educational resource network.

- youngagrarians.org

BC Festivals

Fibreswest (March), Vancouver

100 Mile Fleece and Fibre Fest (May), Coombs

North Island Fleece & Fibre Fest (May), Campbell River

Salt Spring Island Fibre Fair (July), Salt Spring Island

Fibrations (August), Victoria

Woolith Faire (Spring), Vancouver

Cowichan Valley Fleece and Fibre Fest (October), Cobble Hill

Knit City (September), Vancouver

Weavers & More Sale (October), Duncan

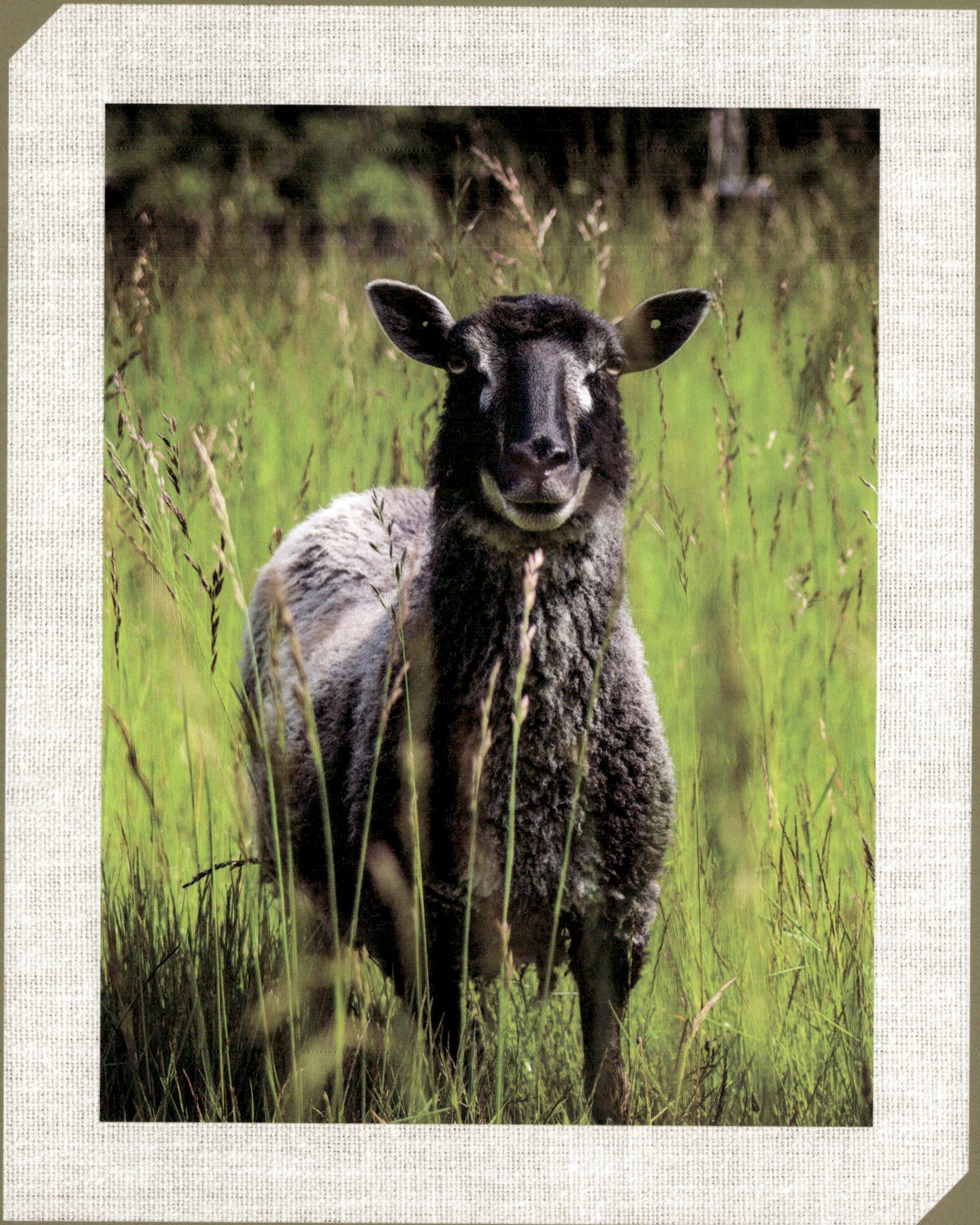

References

◀ A Shetland ewe at Rootspell Farm.

I GATHERED STATISTICS, historical facts, and breed-specific information from many print publications and online resources. This is a comprehensive list of all the resources I most often consulted during my writing. Anyone seeking more understanding about the fibreshed concept, wool history in Canada, or our current world textile economy will find resources here to broaden their understanding.

Burgess, Rebecca and Courtney White. *Fibershed: Growing a Movement of Farmers, Fashion Activists, and Makers for a New Textile Economy.* White River, VT: Chelsea Green Publishing, 2019.

Ekarius, Carol and Robson, Deborah. *The Fleece & Fiber Sourcebook: More than 200 Fibers from Animal to Spun Yarn.* North Adams, MA: Storey Publishing, 2011.

Government of Canada (website). "Plastics challenge: Textiles and microfibers." Innovation, Science and Economic Development Canada, accessed February 12, 2020, https://ised-isde.canada.ca/site/innovative-solutions-canada/en/plastics-challenge-textiles-and-microfibers.

Heinrich, Linda. *Linen from flax seed to woven cloth.* Atglen, PA: Schiffer Publishing, Ltd, 2010.

Johnstone Grimmer, Barbara. "Local lamb thriving on BC Islands." *Sheep Canada Magazine* Vol. 30 (2015).

Kallis, S. *Common Threads: Weaving Community through Collaborative Eco-Art.* Gabriola Island, BC: New Society Publishers, 2014.

Olsen, Sylvia. *Working with Wool: A Coast Salish Legacy & the Cowichan Sweater.* Winlaw, BC: Sono Nis Press, 2010.

Ranson, R. *Homegrown Linen: Transforming Flaxseed into Fibre.* Victoria, BC: Crowing Hen Farm, 2019.

Vesley, J.A. *Wool Production in Canada.* Ottawa, ON: Communications Branch Agriculture Canada, 1984.

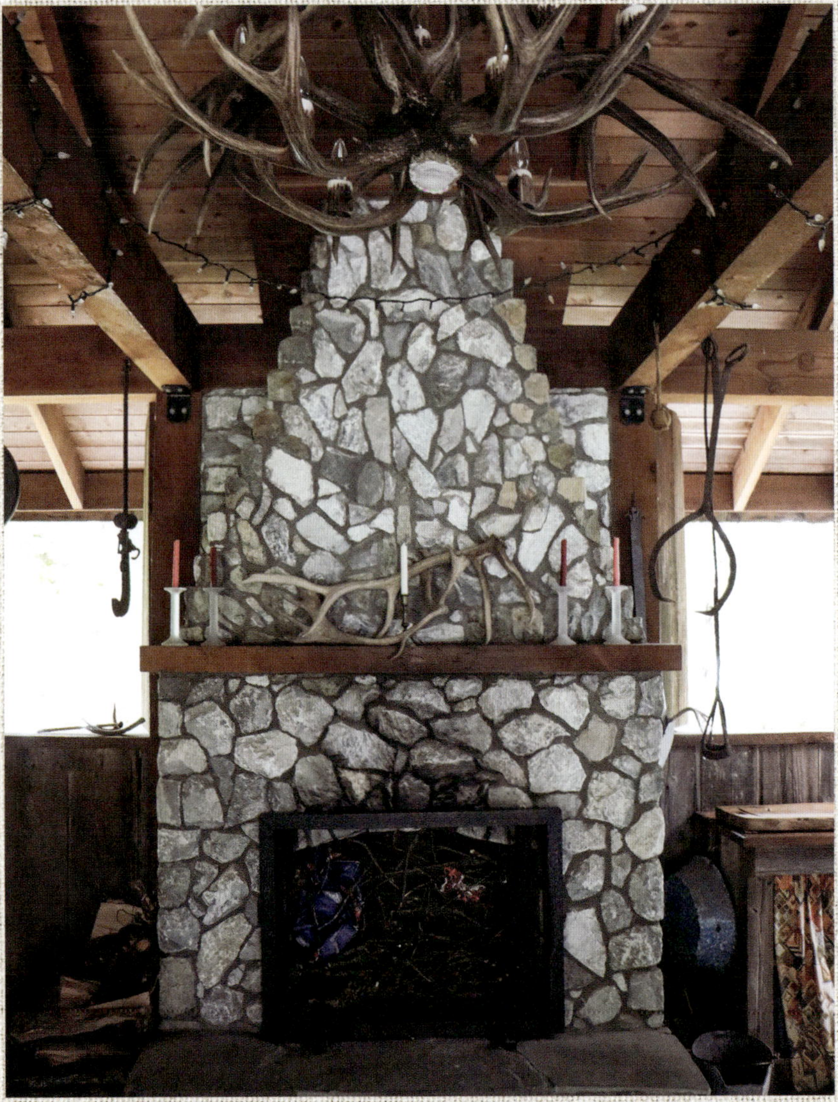

Acknowledgements

◀ The fireplace in the outdoor dining area at Genesta Farm.

FIRSTLY, I WOULD like to acknowledge that this book was written on the unceded and unsurrendered territory of the Stz'uminus First Nation. As an Anishinaabe woman I am not originally from Vancouver Island but I am forever grateful to call this place home.

This book is an offering of gratitude to the farmers, fibre processors, and makers of Vancouver Island and the Gulf Islands for all the hard work you do to bring local fibre to our market. For those of you who support our local fibre economy by purchasing local or supporting organizations like fibreshed—thank you.

Thank you to Deborah and Don Wytinck for your hospitality and for staying in touch. Thank you to Hanahlie Beise for feeding me the best zucchini fritters I've ever had and letting me hang out with your beautiful family. Thank you to Anne Guthrie for offering to help me with my delinquent border collie.

Many thanks to Lynda Drury for all your hard work with the Vancouver Island Fibreshed and for all the valuable information and local fibre knowledge you passed along to me years ago.

While writing this book, I had many questions arise about publishing. Joy Gugeler and Jay Rusesky, professors at Vancouver Island University, were always there to help me out—thank you.

A big thank you to Emily McIvor for reading my writing and giving me valuable feedback when I could no longer see my work.

▶ Icelandic fleece.

Kate Gateley, your tenacity inspires me. I am forever grateful to have you in my world.

Codie Jones and Danielle Witherspoon, thank you for being you. You're those people who always bring me back to myself. You remind me of who I am and what I am capable of. I love you.

A heartfelt thank you to my partner, Travis Hird. For being the rock. And always feeding me when I forget to eat. There were some weeks when you did it all and I am beyond grateful. Thank you to my sons, Naden and Evan, for your constant love and for taking my writing seriously. You mean the world to me.

Thank you to my Grandmother Renee for always having your own fibre space, even if it was just a corner in your room dedicated to knitting and sewing, and for beginning my love for all things fibre. You inspire me in every way.

To my late Grandfather Francis, for teaching me to get my hands dirty in the garden and for instilling in me a deep love and appreciation for nature.

Thank you to my parents, Tamara and Joe, for your constant love and support. And for teaching me to cherish the little things.

To my extended family and friends who are always cheering me on—thank you.

My work on this book would not have been possible without the financial support from the Canada Council for the Arts. I deeply appreciate the support.

And of course, thank you to the team at Heritage House Publishing for taking a chance on me and turning my dream into a tangible reality. Marial Shea, I don't think I could have been matched with a better editor. Thank you for bringing my work to the next level.

Miigwech. Thanks.